W9-CDX-021

Pressed for Time

Pressed for Time

The Acceleration of Life in Digital Capitalism

Judy Wajcman

The University of Chicago Press
Chicago and London

Judy Wajcman is the Anthony Giddens Professor of Sociology at the London School of Economics, the author of *TechnoFeminism*, and the coauthor of *The Social Shaping of Technology* and *The Politics of Working Life*.

The University of Chicago Press, Chicago 60637
The University of Chicago Press, Ltd., London
© 2015 by The University of Chicago

All rights reserved. Published 2015.
Printed in the United States of America

24 23 22 21 20 19 18 17 16 15 1 2 3 4 5

ISBN-13: 978-0-226-19647-3 (cloth)
ISBN-13: 978-0-226-19650-3 (e-book)
DOI: 10.7208/chicago/9780226196503.001.0001

Library of Congress Cataloging-in-Publication Data
Wajcman, Judy, author.
Pressed for time : the acceleration of life in digital capitalism / Judy Wajcman.
pages ; cm
Includes index.
ISBN 978-0-226-19647-3 (cloth : alk. paper)—ISBN 978-0-226-19650-3 (e-book)
1. Time—Sociological aspects. 2. Time pressure. 3. Time perception—Social aspects. 4. Information technology—Social aspects. 5. Technological innovations—Social aspects. I. Title.
HM656.W35 2015
304.2'37—dc23
2014010889

♾ This paper meets the requirements of ANSI/NISO Z39.48-1992 (Permanence of Paper)

Contents

Preface

Many moons ago in the summer before I started university, I went on a trip to Papua New Guinea, then still an Australian protectorate. I had a pen friend from a coastal fishing village, an hour's slow boat trip from Lae. I remember this time vividly, as I felt like I had experienced what was then thought of as "primitive communism," a utopian natural state of cooperation.

Karl Marx and Friedrich Engels use this concept to denote an early form of society whose members collectively share all resources and in which there is no hierarchy. I have since read "The Original Affluent Society," where anthropologist Marshall Sahlins describes how hunter-gatherers, with very little of what we would call technology, spend most of their time relaxing or in leisure pursuits, as they satisfied their needs with a three- to five-hour working day.

In the village, I was plagued by mosquitoes but tropical fruit grew abundantly for the picking, taro and sweet potato were cultivated in gardens, and fish in the sea were plentiful. The living conditions were indeed primitive, and I slept on a bench next to a row of girls in a women's hut with no privacy whatsoever. But there seemed to be all the time in the world. Needless to say, there were no clocks!

I recall one day, as it was nearing Christmas, spending an entire day making coconut milk. Young and ignorant as I was, I suggested that if we altered the method by which the coconut flesh was squeezed in the cloth, we could make the milk much faster and increase productivity. The young people I was doing this with looked quizzical and told me that the making of the milk always took a whole day. They were not in a hurry. They were not interested in speeding up the process—it always had and always would take a full day to make. Now, forty years later, as I write this book on time pressure, I wonder why I so insisted that saving time should be our overriding orientation, an unquestioned good. I have since learned that the important things in life can't be quantified, timed, measured, or accelerated.

Acknowledgments

Every book builds on an author's previous research, and compiling acknowledgments becomes more difficult with every new project. For decades, many people have helped clarify my thinking about the subjects treated here. I cannot begin to name them all. I owe a general debt to the Society for Social Studies of Science (4S) for providing a stimulating climate for discussion, a wealth of excellent papers at its annual meetings, and the most generous international community of scholars one could hope for.

The London School of Economics and Political Science is the ideal intellectual milieu for a social scientist. I have enjoyed fruitful exchanges with my colleagues here and have learned from my bright graduate students. My only regret is that because of Dave Frisby's untimely death, I could not enjoy more of his insights on Simmel. Several postgraduate students helped with bibliographic work, but I would especially like to acknowledge the fine editorial work of Des Fitzgerald. I am grateful to Louise Fisher for making Sociology a convivial department, and to Paul Gilroy for acquainting me with some aspects of digital culture.

While on sabbatical, I enjoyed the hospitality of several other universities. The Oxford Internet Institute has provided me with a warm home and various affiliations over the last decade. I ran a workshop on time and technology with Leslie Haddon early on in the conception of this book. Thanks to Bill Dutton for initially welcoming me there. Thanks also to Nuffield College, Oxford, where I had the pleasure of being a visiting fellow for a term, and in particular to Duncan Gallie and Jonathan Gershuny. I spent another term at the Institute for Public Knowledge at New York University and thank Craig Calhoun for this opportunity.

My interest in time-use studies was initially stimulated by Michael Bittman, and, together with Paul Jones, we undertook a major research project, "The Impact of the Mobile Phone on Work/Life Balance" at the instigation of the Australian Academy of Social Sciences. My home institution then, the Research School of Social Sciences at the Australian National University, allowed me to devote myself to this project. I am grateful to them all. Emily Rose worked as a brilliant researcher on that project. Some of this research is reported here in chapters 4 and 6. I thank Paul for enriching conversations about the interface between media research and science and technology studies. I would like to credit Philip Sayer for the photographs that appear in the book.

Perhaps inevitably, I have to report that the writing of this book was not accelerated by digitalization. Some old and newer friends who have patiently encouraged me in this project need mentioning: Lynn Jamieson, Donald MacKenzie, Lucy Suchman, Ingrid Erikson, Martha Poon, and Tony Giddens. Jenny Earle's wonderful style is, I hope, evident in some of the writing, and finally, I am delighted to acknowledge my son, Samuel Earle, who never seems pressed for time.

Introduction

Tools for Time

Time, it seems, is at a premium. There is a widespread perception that life these days is faster than it used to be. We hear constant laments that we live too fast, that time is scarce, that the pace of life is spiraling out of our control. Phrases such as "high-speed society," "acceleration society," "time famine," and "runaway world" portray more and more aspects of our lives as speeding up.

These concerns are reflected in debates about work-life balance as people try to cope with the pressures of contemporary society, finding enough time for work, time for families, and time for leisure—even time for sleep. Indeed, the desire to slow down the pace of life increasingly features in studies of happiness and well-being. A lack of control over one's time and unequal access to leisure are being identified as important dimensions of social justice. As leading European social scientist Helga Nowotny summed it up in her classic book *Time*, the challenge for modern citizens, who are liable to feel increasingly harried, is to "find time for themselves."[1]

But hang on a moment. Weren't modern machines supposed to save, and thereby free up, more time? Not so long ago, commentaries about postindustrial society predicted a "leisure revolution" driven by automation in industry and the home. Economic progress and increased prosperity would free people from having to focus on providing for subsistence needs, delivering more leisure time. Sociologists talked of the "end of work" and wondered with some apprehension how the vacant hours would be occupied.

Instead, the iconic image that abounds is that of the frenetic, technologically tethered, iPhone- or iPad-addicted citizen. Academic discussions of the impact of digital devices, such as the Internet and the smartphone, typically confirm the popular view that technologies are speeding up life and making us busier. Rapidly evolving information and communication technologies are seen as marking a whole new epoch in the human condition. It is as if the exponential growth in computing power predicted by Moore's law applies to every aspect of modern society.[2]

As technologies proliferate, we find that we do not have more time to ourselves; in fact, many of us have less. How, exactly, has technology hastened the pace of everyday life? How has it made us busier rather than making us more free? Why do we turn to digital devices to alleviate time pressure and yet blame them for driving it? This is the central paradox that I want to examine in this book.

Modern patterns of time can scarcely be conceived of without the use of technology. We rarely have a chance to live outside technologies—they are inextricably woven into the fabric of our lives, from birth to death, at home, in school, in paid work, and in leisure. From simple tools to large technological systems, our lives are intertwined with technology. We delegate tasks to devices and use them to mediate ever more complex social networks. Our actions and society itself are built on and with technical artifacts.

While sociologists emphasize that time is a social entity, formed through collective rhythms of human engagement with the world, technology is rarely accorded the same treatment.[3] Technology is too often seen as outside social relations. But if time cannot be separated from the collective rhythms, assumptions, and hopes of human life, then neither can the technologies that increasingly mark and shape time for us. This may not have been an important distinction in previous eras—but in the digital age, it really matters. For example, the tyranny of the clock, with its linear measurement of the hours of the day, is basic to narratives of the accelerating world. It is as if technical devices incorporate functional time demands that determine unequivocally our uses of time.

If we have, up to now, been too quick to accept the temporal logic built into our new technologies, then we need to remember that the social determination of time is built in to those technologies, just like the size of the screen is, or the power of the processor. Consider, for example, the fiber-optic cable between Chicago and New York. While previous cables between the two cities had been laid along railway lines, the new cable takes the shortest route possible, even drilling through the Allegheny Mountains. It shaves 1.3 milliseconds off the transmission time of the earlier cables. "Speed" is thus built directly into the design: the cable was laid where it was to make transmission faster. But what compels its use by financial trading firms isn't anything directly technical; rather, it is the structure of competition among such firms.[4] Temporal demands are not inherent to technology. They are built into our devices by all-too-human schemes and desires.

This is the thesis at the heart of the book. It enables us to leave behind the old dichotomies about technologies being either inherently liberating or enslaving. By now we should have learnt to be skeptical about both extreme positions: the messianic promise of a technologically-wrought new epoch on the one hand and a blanket rejection of dominance by machines on the other. The digital world is neither exactly the same nor completely different from the industrial world. In order to understand our current obsession with speed, we would be better off exploring *both* the things that have stayed the same, and the things that are particular to our time.

To this end, a historical sense of "new" technologies is required. Machines in the industrial age recast people's experience of time, just as they continue to do now. Yet considerations of technology's impact on time fixate on the latest gadgets, while older dependable artifacts are so familiar as to be left forgotten in the shadows. I will call into question this implicit division between cutting-edge technologies and existing technologies, the spectacular and the ordinary. With this in mind, we will be less inclined to identify technology itself as the source of positive or negative change.

It is our own concrete social practices that generate those qualities of technologies we usually consider as intrinsic and permanent. In other words, technologies only come to life and have meaning as people adopt and use them. At the same time, technologies play a central role in the constitution of time regimes, as our very experience of human action and the material world is mediated by technology. It is simply impossible to disentangle our notion of time from our embodied habitual involvement with the sociomaterial world. We make the world together with technology and so it is with time.

My interest throughout, then, is in exploring the mutual shaping or coevolution of new technologies and temporal rhythms. Broadly speaking, the book takes a *social shaping* approach to technology, regarding technological change as open-ended and unpredictable, but shaped by a range of social, economic, and political forces.[5] Much work in this vein has investigated the design or materiality of specific technologies, but this book has a rather different focus. While recognizing that devices are designed with particular capacities or affordances, I argue that there is nothing inevitable about the way they evolve and are used. Their relation to time depends on how artifacts enter and become embedded in our institutions and the quotidian patterns of daily life—this goes for organizations, user cultures, in production and consumption, family, leisure, and work.

Feeling Harried

Why is there such intense cultural concern with the relationship between digital technologies and the speed of life? That people's subjective sense of time pressure has become urgent is illustrated by the fact that a rising proportion of the population report feeling short of time. Numerous surveys indicate that Americans, as one example, feel more rushed, harried, anxious, and pressed for time than ever before.[6] Psychological and psychiatric diagnoses associated with time strain are also growing. This evidence about the subjective experience of time suggests a widespread perception of everyday life as "time squeezed" and "harried." Leisure time too seems intensified and scarcer.

The objective facts of the matter are, however, far from clear. Indeed, attempts to measure if and how time use has changed over the last few decades have uncovered an intriguing conundrum. The consensus among time-use researchers is that leisure time has, if anything, increased. While Americans complain ever more stridently that they are overworked, estimated average weekly employment hours changed minimally between the 1970s and the 2010s. And if you aggregate the total amount of work time (paid and unpaid), the figure of five hundred minutes a day, or just over eight hours, has remained more or less stable over the last fifty years.[7] Add to this the fact that we are on average living longer, so have more years at our disposal, and this mismatch between objective and subjective time becomes even more intriguing.

This lack of congruence between the amount of free, discretionary time available to us and our contemporary feelings of harriedness has become known as the *time-pressure paradox*. In this book I will expose some of the underlying myths and misconceptions about our high-speed society.

An important starting point is to recognize that there is no single story about what is happening to the tempo of people's lives. Talking in terms of statistical averages masks major shifts in different directions for diverse groups whose time use is being aggregated. For example, as income inequality increases and working hours polarize, discretionary time also becomes more unevenly distributed (although not necessarily in the way one would expect, as we shall see). How much time one has and how it is apportioned can only be understood as a function of underlying social and economic patterns. And, rather than time pressure being an individual phenomenon, I will show that it is related to transformations in household composition and gender relations over the last few decades.

A more fundamental limitation of discussions to date is the narrow concentration on the *quantity* of time available. This is the only aspect of time that time-use studies measure. In order to explain the gulf between so-called objective and subjective time, however, we need a more nuanced understanding of the *quality* or character of time. Time poverty cannot simply be understood in terms of the amount of (clock) time available. How people experience and practice time is the result of the meanings and values that they ascribe to various kinds of activities. The demand for "quality time" with children is a popular expression of this. Not all activities are performed at the same pace, nor would we want them to be. The result is conflicting temporal regimes that require coordination, which, in turn, leads to time pressure. Leisure time itself may be subject to intensification because of the increasing habit of multitasking with digital devices.

Harriedness is, thus, a multidimensional phenomenon. There are both different senses of feeling pressed for time and a range of mechanisms that trigger these feelings. I will explore the role that these multiple dynamics play in accounts of the time-pressure paradox.

Mapping the Discussion

The book explores these ideas thematically and is organized in the following way. The first chapter begins by outlining the extent to which social theorists link modern society to processes of acceleration. Here we see that time-space compression is a recurring theme, with the main impetus being information and communication technologies. For some, digitalization is even generating new kinds of *timeless time*, or *instantaneous time*, eclipsing the linear logic of clock time.[8] The phenomenal speed of financial trading is emblematic of this.

But what does acceleration mean and is it a constitutive trait of our era? I argue that a lack of clarity about the concept itself serves to sustain the belief that faster machines catapult us into a faster life. Accordingly, I distinguish between different kinds of acceleration, opening up the connections between them to scrutiny. My particular interest is in how technology affects the pace of everyday life, and I question the determinist role ascribed to it in theories of the high-speed, network society. In contrast to the sweeping, celebratory claims made by cyber-gurus, I suggest that the social studies of technology offer a richer analysis of the reciprocal relationship between time and machines.

Talk about life accelerating only makes sense against an implied backdrop of a slower past. Chapter 2 thus provides a necessary historical perspective. Most sociologists explain our modern sense of time as dating from the commodification of time in industrial capitalism. With the exploitation of labor, saving time becomes equivalent to making profit, as expressed in Benjamin Franklin's famous equation of time and money. These arguments focus on the economy, stressing the way that clock time becomes time, per se, and is internalized as time discipline.

However, it is in the city where speed becomes a general condition of modern life: the antithesis of the slow pace of country life. Beginning around the turn of the nineteenth century, speed itself becomes closely identified with narratives of progress. This period saw dramatic technological inventions, such as the steam engine, railways, the telegraph and the telephone, that transformed people's consciousness of time, space, and their environment.

Georg Simmel's prescient analysis of the ambivalent consequences of the increased pace of metropolitan life is a rich source for this chapter. The symbolic significance and allure of speed that characterize our present condition is shown to have a long lineage. Indeed, cultural values and consumption, as well as production, affect perceptions of time scarcity, as we shall see.

To this point, I have been treating the phenomenon of acceleration as a characteristic of society as a whole. However, as I have already indicated, not everyone has a uniform experience of time. In chapter 3, I dig deeper into the empirical data to reveal how time use differs for diverse social groups (as do relationships with technology, which I consider in later chapters). It is apparent, for example, that time poverty is far greater for single parents than it is for couples without children, and that women tend to have less temporal autonomy than men. Drawing on time-use surveys, the most reliable source we have on the allocation of time, I demonstrate how changes in work patterns, household arrangements, and parenting all affect time stress. A major factor that emerges is the difficulty of arranging shared time with family and friends in a desynchronized society. Finally, picking up a thread from the previous chapter, there is the prestige attached to a busy lifestyle that swells the refrain of relentless haste in some quarters.

I then turn directly to the role that digital technology plays in shaping our experience of time. What kinds of temporal rhythms do people create with new technologies? And what difference does it make that our everyday social situations and communications are increasingly characterized by ubiquitous, multiple modes of connectedness? The rest of the book focuses on this theme, with particular emphasis on "new media," information, and communication technologies.

Working time is the subject of chapter 4. I examine not just the length of working hours but the pace, intensity, or what one might think of as the temporal density of work. I evaluate evidence for the intensification of work and the ongoing impact of information technology in different occupations and industries. The volume of e-mail and the constant connectivity afforded by mobile phones are commonly seen as key causes of stress at work. Office life has become identified with information overload, endless interruptions, multitasking, and raised expectations of speedy response time. I question this stereotype of the technologically tethered worker with no control over his time. Instead, the use of information and communication technology (ICT) for both work-related and personal matters is shown to have positive, as well as negative, implications for men and women workers. The contemporary office has by my account morphed into a ubiquitous technoscape, and this has reconfigured the very nature of working time.

Chapter 5 looks at how households allocate time to unpaid domestic work. Overall, time-use surveys show significant gender differences in the amount of time spent and the type of activities performed. While fathers are doing more and women are doing less than they were, women's labor still accounts for over two-thirds of the total time devoted to unpaid work. Is technology the solution? Will the digital home of the future finally free us from time-consuming drudgery?

This chapter will consider why "time saving" domestic technologies, such as washing machines and microwave ovens, have been surprisingly unsuccessful in lessening the domestic load. It turns out that the effect of technology has been peripheral because of changed expectations about child care, the emergence of new standards and tasks, and the durable connection of domestic activities with masculine and feminine identities. Finally, I examine the cultural imaginaries around smart houses and "caring" robots, showing how they reflect the ethos of their designers.

As well as being the site of housework, the domestic sphere is associated with personal relationships and intimacy. How communication technologies affect these dynamics is the theme of chapter 6. I begin by describing the saturation of everyday life by media technologies, for example, how people are meshing multiple devices. I then present some of my own research on the mobile phone's role in shifting the boundaries between work and home. The main use of the mobile phone turns out to be social, with much value placed on the enhanced ability to microcoordinate the timing of complex family activities. In this way, I argue, mobile phones have become a new tool for intimacy.

More broadly, I consider the effect of the machine-mediated nature of social relationships. The significance of the move from mass communication, such as television and radio, to individual, privatized communication on personal digital devices is contentious. Opinions tend to be polarized between those who stress the new freedoms and increased individual autonomy enabled by these technologies and those who foresee a world of constant connectivity but less meaningful communication. A false dichotomy has developed between direct and mediated communication, positing them as alternatives. I argue, instead, that we need to think about the ways in which communication and affect are embedded in material objects.

The final chapter explores some possible directions for making more of time, like reducing working hours. Such strategies, however, require revision as portable technologies rearticulate what were once distinct boundaries between "my time" and "work time". We see again how the very same devices that can make us feel hurried also free up time. Indeed, I suggest that digitalization provokes a radical rethinking of the standard debates about work-life balance and their oppositional character. I then turn to examine the claim that ICTs have intensified consumption and leisure, what is often described as a culture of instantaneity or immediacy. Once more I show how the story is complicated, with accelerating time frames being offset by the emergence of unforeseen, slow ones. The relationship between technological change and temporality is dialectical, not teleological.

The cult of speed has prompted slow living movements that attempt to alter the tempo of everyday life. The exemplar is slow food, and I discuss both its appeal and shortcomings. In particular, I encourage a critical distance from the notion that rejecting high-tech systems points the way forward. Rather, I argue that ICTs make possible new and multiple temporalities. But will we get the type of technology that is best suited to this endeavor? I consider the extent to which innovation is conflated with efficiency narrowly conceived, particularly by the engineers of Silicon Valley. This shapes how social problems are formulated, the kinds of machinery that are developed, and even our sense of self. Perhaps most perniciously, our visions of the future are profoundly pervaded by their preoccupation with ever more acceleration.

Conclusion

Throughout the book, I will explore many of the complexities and nuances that structure and texture the different ways that people are caught within the webs of time, technology and daily life. Clearly not everyone is caught up in the accelerating dynamism of modernity in the same way. Some of these nuances will be particularly important in what follows, and I pay close attention to the role of gender, to different kinds of labor, and to varying patterns of technology use by age.

Inevitably, my focus on the interconnections between speed, technology, and the relationship between work life and leisure has led me to focus on predominantly Western, "overdeveloped," industrial economies. It has also led me to focus on people who are in work within those economies. Indeed, throughout the book, where possible, I draw on my own empirical research that has largely been based within the work spaces of various Anglo American economies. But if unemployed people and people living in the global south are somewhat missing from this account, still much of what I describe is part of global and more general social trends towards urbanization and technologization.

What I very much want to do here—and what this narrowed focus has allowed me to do—is to bring technology back into the conversation about speed and time. There are numerous theories about fast, mobile capitalism that do focus on technology, but these are rarely informed by how time is actually practiced. It is my intention in this book to bridge this gulf, marrying abstract social theories about modernity and acceleration with a wide range of empirical studies. This involves developing a unique dialogue between several social science literatures that are usually kept distinct.

Only such a wide-ranging discussion will enable us to assess the acceleration thesis. While economic, technological, social structural, and cultural changes in modern societies have altered the experience of time in unprecedented ways, the picture is not uniformly one of speeding up. If we are short of time for work, parenting, friendship, leisure, and civic participation, this is not simply a function of machines, old or new. Technologies in themselves do not lead to either velocity or slowdown.

In the digital age, however, our communication patterns and interpersonal sociability are much more mediated by and distributed across a whole range of multimodal devices. In every sphere, we inhabit a technologically suffused environment in which constant connectivity is the norm. Such interconnected sociomaterial networks are transforming the pace and scope of human interaction. This is giving new meanings to temporality and reconfiguring our time practices. The social shaping perspective elaborated here offers a framework for understanding the myriad ways in which the rhythms of our lives are intertwined with technologies.

Such an analysis has implications for our politics. It means that there is no technical solution for our current condition. We cannot simply go on a digital diet, reject the smartphones and return to nature, as posited in some deceleration arguments. Nor should we look for promises of emancipation in technological futures populated with social robots. These cultural imaginaries themselves reflect the dominant engineering approach to time saving and time ordering. Instead, the process of technical innovation and design needs to be opened up to reflect a broader range of societal realities and concerns. Rather than digital devices pushing us inexorably into a life in the fast lane, I will argue that they can be actively appropriated and recruited as an ally in our quest for time control.

Time may appear inherently egalitarian, in that everyone has just twenty-four hours in a day, seven days a week and twelve months a year, and this will remain the case in the epochs to come. However, temporal sovereignty and sufficient leisure time are important indicators of a good life. How much time we have is both a crucial aspect of freedom and individual autonomy, and a measure of equality. This book is intended as a contribution to understanding the role that technology plays in the time of our lives.

Chapter One

High-Speed Society

Is the Pace of Life Accelerating?

> Any attempt to make sense of the human condition at the start of the new century must begin with the analysis of the social experience of speed.
>
> WILLIAM SCHEUERMAN, *Liberal Democracy and the Social Acceleration of Time*

The relative speed of society has long been seen as one of its essential characteristics. Many of the inventions that are considered pivotal to progress, from the wheel to the microchip, have been designed to enable us to go faster. Yet it is in diagnoses of our contemporary times that acceleration features most prominently. Time-space compression, the idea that technologies have dramatically telescoped temporal and spatial distances, is a constant motif, as is the notion that economic, social, and cultural change is much more rapid than in previous eras. Things seem to happen at a relentless pace, imbuing us with a different sense of time.

According to the dominant narrative, our ubiquitous experience of busyness makes perfect sense as we inhabit a high-speed society. Our age is obsessed with speed: faster cars, faster trains, faster broadband, even speed dating. Speed is sexy, and digital devices are constantly sold to us as efficient, time-saving tools that promote an exciting, action-packed lifestyle. Nowhere is this more apparent than in iPhone's Siri software, which allows you to "use your voice to send messages, schedule meetings, place phone calls, and more," while, the advertisement suggests, driving or exercising. Similarly, self-logging wristbands that track everything from heart rates and sleep patterns to mood fluctuations are marketed for a busy life on the move.

Our obsession with doing more at once is symptomatic of the frenetic pace of life. The yellow brick road may wind through the Googleplex, with all its indoor tree houses, volleyball courts, apiaries, and giant, colored rubber balls, but over-the-rainbow Google engineers talk of needing to work smarter and harder than they could ever have imagined. Although speed and timing is of the essence, Zen masters are brought in to teach employees how to stop and take a deep breath. The typical mantra of CEOs is that technology is pushing us faster and so we have to adapt to new ways of doing business in "a world of screens, texts, cell phones, information all over you."[1]

Like corporate speak, much social science sees technology as the main force driving acceleration. The idea that digitalization has spawned a new temporality is widespread and is variously described as *instantaneous time, timeless time, time-space compression, time-space distanciation, chronoscopic time, pointillist time,* or *network time.*[2] There are even calls for a new science of speed or, as Paul Virilio has termed it, *dromology*. All of these concepts have at their core a view that life is speeding up. The spread of communication technologies in particular and their evident potential for the further speeding up of an already accelerated modernity has added urgency to the question of speed and human reactions to it.

But if acceleration defines our digital universe, what precisely does this mean? Despite the dazzling array of theories depicting the present era as one of exceptional speed, the concept remains vague and elusive. The fact that so much of the academic and popular commentary is prone to speculative hyperbole compounds the problem. This, in turn, is exacerbated by the extent to which the agenda for discussing the future of technology is set by the promoters of new technological products.

I begin this chapter, then, by disentangling the rhetoric in order to clarify the relationship between technological acceleration and the pace of life. I also present an overview of the most influential accounts of high-speed network society, which will help to expose the technological determinism implicit in such theories. Perhaps this is an unfortunate but necessary corollary of the scale and scope of the authors' arguments. What gets downplayed or lost, however, is the extent to which the "virtual" is made up of wires, buildings, and bodies, as well as the fact that real human beings know and use (or not use) information and communication technology (ICT) in concrete, local settings. My approach contrasts these tactics by firmly grounding the discussion of how digital time is perceived, organized and negotiated in commonplace everyday situations.

Moving forward, I principally draw on literature from science and technology studies—STS for short—which, for some time, has been urging a more nuanced understanding of the ways in which technology shapes time. Taking on this lens allows us to see that society is more than its technology, but also that technology is more than its equipment. In other words, the social world cannot be reduced to the technology that makes it up. However, this is not at all to diminish the role of technology—in fact, quite the opposite. It is only by paying attention to sociomaterial practices that we can begin to see the rich interplay of technology and society.

Such an approach necessarily questions all-embracing, linear narratives about everything speeding up, pointing instead to a more complex temporal patterning of experience. It requires us to pose questions, such as when, and where, people encounter accelerations (as well as decelerations) and what the consequences for the quality of our lives are.

Acceleration Society

Although acceleration itself is rarely regarded as the central topic for sociological analysis, it is ever present in theories of contemporary society. Physicists have clear ideas about what speed and velocity mean but, in describing human experiences of time in high-speed society, the term is used to refer to a variety of phenomena. This is particularly confusing, as time compression has multiple dimensions, so that while some aspects of life are accelerating, others may not be and could even be slowing down.

One notable exception is Hartmut Rosa, who examines in detail what it means to say that Western societies are acceleration societies. I find his definition, and the distinction he draws between different aspects of acceleration, helpful and have adopted it here.[3]

The first and most measurable form of acceleration is the speeding up of transport, communication, and production, which can be defined as *technological acceleration*. The second is the *acceleration of social change*, meaning that the rate of societal change is itself accelerating. The central idea here is that institutional stability (in the realms of the family and occupations, for example) is generally on the decline in late modern societies. The third process is the *acceleration of the pace of life*. It is the focus of much discussion about cultural acceleration and the alleged need for deceleration. The pace of (social) life refers to the speed and compression of actions and experiences in everyday life.

Now the most intriguing question is how these three types of acceleration relate to each other. As Rosa points out, there is clearly a paradox between the first and third process. If technological acceleration means that less time is needed (for production, transport, etc.), this should entail an increase in free time, which in turn would slow down the pace of life. Rather than time becoming abundant, however, time seems to be increasingly scarce. Accordingly, it only makes sense to apply the term *acceleration society* to a society if "technological acceleration and the growing scarcity of time (that is, an acceleration of the 'pace of life') occur simultaneously."[4] Interrogating this "time pressure" paradox is the central quest of my book.

According to this definition, most general analyses of contemporary society can be read as versions of the acceleration society thesis. In other words, they make a direct, causal link between technological acceleration, especially the speed of electronic communication systems, and the harriedness of everyday life. The fact that our social interactions in both work and leisure time are increasingly mediated by technology—that we live in a state of constant connectivity—is a recurring theme. Here I want to focus primarily on how the connection between the speed of technology and the pace of life is formulated.

There is a vast literature on what is commonly referred to as *time-space compression*. Geographer David Harvey classically conceived of this process as being at the heart of modernity, or, in some formulations, postmodernity: "I use the term 'compression' because . . . the history of capitalism has been characterized by speed-up in the pace of life, while . . . space appears to shrink to a 'global village.'"[5]

Key to Harvey's work on the spatial-temporal dynamics of capitalism is the notion that economic processes are accelerating. For him, the driving forces behind social acceleration are globalization and innovations in ICT that facilitate the fast turnover of capital across the globe. In contrast to industrial capitalism, which requires the exploitation of labor through strict adherence to clock time and Fordist spatial models like the assembly line, *flexible accumulation* requires a shift in the ways we think about time. Harvey observes that the general speed-up in the turnover time of capital accentuates the volatility and ephemerality of commodities and capital. Fast capitalism annihilates space and time. The distances that once hampered global trade are made meaningless as humans increasingly communicate using "real-time" technologies. Time becomes beyond control as distance disappears in a world of instantaneous and simultaneous events. Acceleration, then, is reflected in the substantive temporalities of human existence, in particular, the growing sense of time-space compression in everyday life.

Such discussions of acceleration typically invoke Karl Marx's analysis of capitalism and the constant need to speed up the circulation of capital. The faster that money can be turned into the production of goods and services, the greater the power of capital to expand or valorize itself. With capitalism, time is literally money, and "when time is money, then faster means better" and speed becomes an unquestioned and unquestionable good.[6] Technological innovations play a key role in that improvements in the conveyance of communication, commodities, and bodies reduce the cost and time of capital circulation across the globe (what Marx called the "annihilation of space by time"). The extent to which such time-space compression would be fulfilled, however, was unforeseen by Marx.

Developments in the speed of transport and communications have shrunk the globe, from the horse-drawn coach and sailing ship to jet aircraft today. Yet it was only with the invention of the telegraph in the 1830s that the carriage of bodies by wheel, sail, and steam was challenged by the transport of messages at speeds dramatically different from those that had previously existed. The telegraph meant that a message could be delivered at a tiny fraction of the time afforded by physical transport.

Electronic communication has increased this speed in exponential ways. The velocity of automated financial trading, which is now moving from milliseconds to microseconds (millionths of a second), has become emblematic. This is far faster than human reaction times, which typically range from around 140 milliseconds for auditory stimuli to 200 milliseconds for visual stimuli. In this context, even a 5-second pause can seem like a very long time.[7] Indeed, the exponential growth in Internet transmission speeds over the last 100 years is accelerating to the point where data can be transferred at a sustained rate of 186 gigabits per second, a rate equivalent of moving 2 million gigabytes in a single day.[8]

Our own sense of time has been profoundly altered by the convergence of telephony, computing, and broadcasting technologies into a pervasive environment of instant and simultaneous information and communication. So it is not so surprising that, in the face of such an intense phase of time-space compression, and the resulting changes in our time consciousness, many social scientists herald a new social order.

The problem, as I will show, is that theories about social acceleration are too schematic to capture the multiple temporal landscapes, both fast and slow, that come into play with digital devices. The prose is all about "virtual" networks and ubiquitous computing, which are conceived of as borderless disembodied spaces and ethereal instantaneous times. This has the effect of rendering invisible the tangible human and social time dimensions of everyday life as "banal, repetitive, and trivial."[9] In other words, the quotidian time of intersubjectivity, in which actual women and men coordinate their time practices in real-world contexts, gets completely obscured.

The Network Society

Perhaps the best-known example is Manuel Castells's *The Rise of the Network Society*. For him, the revolution in ICT has given rise to a new information age, a network society in which labor and capital are replaced by informational networks and knowledge. Information is the key ingredient of organizations and flows of electronic messages and images between networks now constitute the basic thread of social structure. He defines the space of flows as the technological and organizational possibility of practicing simultaneity without contiguity. These circuits come to dominate the organization of activity in individual places such that the site of networks and their relationship to other networks become more important than the characteristics of place itself. For Castells, the information age, in which virtuality becomes an essential dimension of our reality, marks a whole new epoch in the human experience.

For our purposes, here, what is particularly interesting is his argument about the disappearance of time: that we are increasingly moving away from the clock time of the industrial age, in which time was a method of demarcating and ordering sequences of events.[10] Instead, he argues, the world is increasingly organized in the space of flows: flows of merchandise, people, money, and information around dispersed and distributed networks. The sheer velocity and intensity of these global flows, interactions, and networks dissolve time, resulting in simultaneity and instant communications—what he terms *timeless time*. While this new timeless time emerged in financial markets, it is spreading to every realm. No wonder then, Castells opines, that life is a frantic race as people multitask and multilive by means of technology to reach "timeless time: the social practice that aims at negating sequence to install ourselves in perennial simultaneity and simultaneous ubiquity."[11] In true postmodern rhetoric, society becomes eternally ephemeral as space and time are radically compressed to the point where, at least with regard to the latter, it ceases to exist.[12]

This vision of the network society, in which the accelerating speed of ICT annihilates time, has been extremely influential. For instance, echoing Castells's concept of timeless time, John Urry argues that new technologies generate new kinds of *instantaneous time*, characterized by unpredictable change and quantum simultaneity. This new time is based on inconceivably brief instants that are wholly beyond human consciousness and, as a result, the simultaneous character of social and technical relationships replace the linear logic of clock time. According to Urry, instantaneous time is also a metaphor for the widespread significance of exceptionally short-term and fragmented time.

While such conceptions of time do capture something important about the extent to which the extraordinary speed of technologies is transforming the economy, financial markets, politics, and patterns of production and consumption, it is far less clear what this speeding up means for the experience of lived time. Urry does include in his specification of instantaneous time "the sense that the 'pace of life' throughout the world has got too fast and is in contradiction with other aspects of human experience."[13] The tenor of his discussion of instantaneous time is that it is socially destructive, yet he does not provide systematic empirical research to support this claim. One is left wondering what time "organised at a speed that is beyond the feasible realm of human consciousness" might mean to people and how it concretely relates to the actual use of ICT in everyday life.

Let me provide just two brief examples. Surely highly mobile, "hot-desking" professionals would be a good test of the notion of timeless time, as their spatial-temporal practices are fundamentally altered. Yet according to a detailed study, rather than time disappearing, their time became dominated by a concern to connect in time and space because they considered face-to-face meetings to be the paramount means of communicating in organizations.[14] As a result, one of their main uses of asynchronous technologies (such as voice mail and e-mail) was to make arrangements for synchronous communications. That the digital media industry is so geographically clustered in both London and New York similarly attests to the importance of "live" social networking.[15] In this sense, local time is hardly superseded. My own research on the contemporary workplace, detailed in chapter 4, shows that while network technologies do alter the tempo of work, the myriad ways in which people deploy their devices can hardly be described as the annihilation of time.

Or take the extreme case of time-space compression, finance. Even here we do not find Castells's immaterial world where time, place, and bodies are replaced by virtual information networks. Financial trading is in fact underpinned by materiality: physical, technological, and corporeal in nature. Trading centers are large warehouses, consuming vast amounts of electric power to dissipate the heat generated by fast computing. There are relatively few staff but rows and rows of computer servers and digital switches and miles of cabling connect those servers to the matching engines and the outside world. By today's standards, a very large data center might be a five-hundred-thousand-square-foot building demanding fifty megawatts of power, which is about how much it takes to light a small city. To guard against a power failure, they further rely on banks of generators that emit polluting diesel exhaust. The ethereal imagery of virtual data stored in the "cloud" is belied by the brute physicality of the infrastructure it needs.[16]

Moreover, contrary to perceived wisdom, the ultrafast reaction time actually increases the importance of spatial distance. It turns out that high-frequency trading firms rent space for their computer servers in the same building as an exchange's engines precisely because the obdurate physical reality of colocation is still important. Time advantages of tens of microseconds become a crucial issue for traders. And these same technologies, working through different institutional arrangements, shape trading in very different ways.[17] "For all the breathless talk of the supreme placelessness of our new digital age, when you pull back the curtain, the networks of the Internet are as fixed in real, physical places as any railroad or telephone system ever was."[18]

While both Castells and Urry explicitly distance themselves from a technologically determinist stance, they do not entirely succeed. At times they have a tendency to discuss the impact of electronic information systems as having major "irreversible" effects, ushering in disruptive social revolutions. The idea that technical innovation is the most important cause of social change permeates Castells's analysis of the network society. Reflecting a common tendency in the literature on digital technologies, he argues from extremes, assuming that technologies are used in a uniform way overall and everywhere, revolutionizing work, leisure, education, family relationships, and personal identities.

Ironically, this is a form of technological determinism that suffers from a lack of interest in technology, what it is really made up of, what it consists of, and so on. What I am trying to show in this chapter is that it is precisely by focusing on technology that we can see how technology is implicated in social relations, human interests, history, and culture.

Such commentaries frame the present era as one in which the world is experiencing historically unprecedented change. Yet even a cursory glance at earlier periods of rapid technological change reveals that similar claims were made about their overpowering effects. In the late nineteenth century, for example, Anglo American culture was fascinated by the capacities of the telegraph and telephone to extend messages effortlessly and instantaneously, annihilating space with time. Indeed, the idea that inventors were ahead of their time and that science and technology were advancing faster than the ability of human society to cope was commonplace. As the next chapter expounds, a sense of increasing speed or acceleration has been a central feature of social commentaries since at least the nineteenth century.

Furthermore, detailed histories of technology immediately suggest that technologies have divergent effects, operating in different ways for different people at different periods in history. What Carolyn Marvin terms *instrument-centered perspectives,* in which the instrument determines the effect, are much too narrow, because even the history of electronic technologies

> is less the evolution of technical efficiencies in communication than a series of arenas for negotiating issues crucial to the conduct of social life; among them, who is inside and outside, who may speak, who may not, and who has authority and may be believed. Changes in the speed, capacity, and performance of communications devices tell us little about these questions. At best, they provide a cover of functional meanings beneath which social meanings can elaborate themselves undisturbed.[19]

My intention is to examine precisely how digital technologies are reshaping our sense of time without succumbing to the common obsession with novelty. As a keen observer of technical processes, I am skeptical about overarching claims in the form of grand, totalizing narratives of postindustrial, information, postmodern, network society. Such theories tend to take the form of "techno-epics heralding techno-epochs" and treat time as an epiphenomenon with relatively little substantive content.[20] While I would not diminish for a moment the importance of social theory, my feminist sensibility also attunes me to the situated and contingent character of truth/knowledge claims and the need to beware of the "god trick."[21]

As I have already intimated, we can best advance our understanding of the dynamics of acceleration through scholarship that is specific, empirical, and located in concrete social settings where those effects can be most clearly observed. I have therefore chosen, in the chapters that follow, to concentrate on how technologies shape our practical perceptions, ideas, and experiences of social time in the unheroic sites of ordinary life. Doing so enables us to investigate the full spectrum of positive and negative consequences of the increased pace of life in modernity, the extent to which it is occurring, and the uneven distribution of these processes.

The Nihilism of Speed

It is worth first pausing to reflect on the work of the French philosopher Paul Virilio, for whom issues of speed and technology are pivotal. His *dro-*

mology, from the Greek word for race (*dromo*), is a theory of the nature of speed, its conditions of emergence, its transformations, and its effects. For Virilio, "speed, the cult of speed, is the propaganda of progress" and its consequences are devastating. "Today we are entering a space which is speed-space. . . . This new other time is that of electronic transmission, of high-tech machines, and therefore, man is present in this sort of time, not via his physical presence, but via programming."[22] If there is fear, he tells us, it is because the earth is shrinking and space is dwindling, compressed by instantaneous time. Carried along by the headlong rush of an increasingly accelerated world, all we can do is manage and administer this fear instead of fundamentally dealing with it. "Climate chaos, stock market panics, food scares, pandemic threats, economic crashes, congenital anxiety, existential dread"—yet we are all still convinced that more speed and ubiquity are the answer.[23] By this account, then, speed is nihilism in practice.

Refreshingly, although Virilio is known as the "high priest of speed," he argues that speeding up is not unique to the digital age.[24] Rather, he suggests that we can read the history of modernity as a series of innovations in ever-increasing time compression. His analysis of speed encompasses nineteenth-century *transport* (trains, cars, and airplanes) that dramatically shortened travelling time, twentieth-century *transmission* (the telegraph, telephone, radio, and computer and satellite communications) that have replaced succession and duration with simultaneity and instantaneity, and *transplantation* that compresses time by providing xenotransplantation and nanotechnology. Each of these technological innovations enhances the independence of the social relations of time from space and the body.

Although Virilio's concern is to identify broad societal trends, he is highly attuned to contradictions and countervailing tendencies, unlike the theorists referred to above. For example, new modes of transport massively compress the time of travel but also lead to standing still in traffic jams in big cities.[25] Endless queues in crowded lobbies are characteristic of travel by plane, delays and cancellations an integral part of commuting by train. Virilio's *dromological* law, which states that increase in speed increases the potential for gridlock, seems more and more apt. He is also aware that political conflicts may ensue, because acceleration affects different individuals, groups, and classes in uneven ways. For instance, traffic jams and waiting times do not have the same impact on everyone, as the money-rich-but time-poor can use their wealth to purchase speed.[26]

Twentieth-century telecommunications further compress duration to zero. However, once again Virilio notes the tendency for acceleration to produce new forms of deceleration, a recurring theme of mine. *Chronoscopic* time, the intensive (electronic) instant, leads to an overload of information so extensive that taking advantage of only the tiniest fraction of it not only blows apart "real-time" communication but also slows down operators to the point where they lose themselves in the eternity of electronically networked information, a "black hole of globalized interconnectivity." The actual capacity for parallel absorption of knowledge is hugely disappointing. Ironically, the same electronic capacity to be both here and elsewhere in the time of nowhere has brought the body to a standstill. While Castells and Urry, for example, emphasize the mobility and fluidity of people in network society, Virilio recognizes that, at the level of everyday life, people are increasingly stationary—sitting in front of a screen. The chronoscopic time of the ICT revolution—a temporality of instantaneous and continuous connectivity—is, paradoxically, accompanied by new forms of inertia.

Finally, in terms of the twenty-first century, Virilio identifies the time compression afforded by transplantation primarily with prostheses provided by xenotransplantation and nanotechnology. Technological time has moved from the vastness of planetary and earthly space to the micro spaces of organs and cells to what he calls "the heart of the living." In a speculative mood, he writes that artificial rhythms replace natural ones, to be speeded up at will and paced to the dictates of the prosthetic machine. Here again he is prescient in foreseeing that the genetic modification of humans and animals raises moral and ethical issues not anticipated and in advance of regulatory frameworks. Indeed, some have speculated that the point of these technologies is to transcend a biological sense of time, in other words, to arrest time.

Human history, then, can be understood in terms of a race with time, of ever-increasing speeds that transcend humans' biological capacity. According to Virilio, the forces of technoscience are speeding up the world to such a degree that things, even reality, are starting to disappear. Technological time is no longer part of chronological time; we have to conceptualize it as chronoscopic time. This new time encompasses the dead time of travel, the intensive time of electronic connectivity, and scarce time—as immense acceleration leads everywhere to a shortening of time limits and time to think. For individuals as well as society, this transformation in the space-time structure has fundamentally disruptive consequences. The technologies of speed bring about a "derangement of the senses" whereby real space is replaced by decontextualized "real-time" processes and intensity takes over from extensity. As humans cannot possibly absorb this overload of parallel information sources, Virilio calls for a cultural slow-down to protect against the further invasion of technology into lived human experience.

Virilio's highly original vision of the world provides a healthy antidote to overly economistic analyses of contemporary capitalism. He is rare among social theorists in giving war and military technology due weight, identifying the key role of rapid movement in military power as well as the significance of the military for the development of technology.[27] However, in the end, his military paradigm overwhelms all other modalities and experiences of speed. We are left with a rather one-sided account of a world now out of control and a reductionist view of modern scientific knowledge.[28] Perhaps this partially accounts for his dire prognosis that the ceaseless increase in acceleration is leading to nothing other than the "liquidation of the world."

Such cultural pessimism besets much writing about the effects of digital technology on the contemporary contours of time. Barbara Adam, Britain's leading social theorist of time, writes in a similar vein:

> Control is lost due to massively increased speed, instantaneity, and networked connections. Instantaneity means "real-time" processes across the globe coupled with the elimination of linear cause-and-effect relations (in a context of continued linearity). This brings with it loss of time to reflect and act in the intervening period between cause and effect. It turns masters into slaves, designers and operators into the weakest links.[29]

So too does Robert Hassan, who argues that digitally compressed network time, oriented toward pure speed, colonizes all other realms of life, leaving no time for reading, reflection, and resistance. It even robs us of sleep, according to Jonathan Crary's *24/7*, down from eight hours a generation ago to approximately six and a half hours for the average North American adult.[30] Continuous connectivity begets more connectivity, as the devices used to coordinate the constant inflow of networked stimuli actually have the effect of stretching the individual even more taughtly over time and space. "The more we become connected and dependent upon interconnectivity in our jobs and other aspects of our lives, the more we will live in an accelerated mode."[31] Accordingly, this ever-quickening quest for speed becomes a pathology, an inescapable addiction, driving us (like Alice in Wonderland) to run ever faster just in order to stand still.

These authors correctly identify the profound dangers posed by developments in technoscience, such as instruments for control and surveillance, and the penetration of informatics and bioscience into every corner of our lives, including, literally, our bodies. However, they only see the dark side. For example, Hassan's depiction of the accelerated "techno-logic" of the network society evokes an image of "a technologically closed system that allows for no real choice or real freedom of technological expression."[32] As such, it unwittingly legitimates a passive, defensive attitude to technological change. While it is evidently true that we live our lives in a technologically infused environment unimaginable only thirty years ago, these societal theories do not provide much detail as to how, why and even if using ICT inexorably leads to the acceleration of everything.

An implicit antipathy to science and technology forecloses appreciation of the scope that digital technologies might afford for control over time, enabling people to have not only more time but time of their choice. This stance is out of step with the widespread recognition that technoscience is a feverishly contested political field. Indeed, the plea for a slow down brings to mind the ecological feminists whom Donna Haraway chided over a quarter of a century ago for wanting to return to nature rather than becoming impure, hybrid cyborgs. A leading critic of technoscience, Haraway insisted on the liberatory potential of science and technology: "The issue is no longer whether to accept or oppose technoscience, but rather how to engage strategically with technoscience while, at the same time, being its chief critic."[33]

In this spirit, I want to argue that a more well-rounded understanding of the relationship between temporality and technology must inform an emancipatory politics of time. This involves the democratization of technoscience, deciding what sort of technologies we want and how we are going to use them. Resisting technological innovation and calling for deceleration or a digital detox is an inadequate intellectual and political response. Indeed, wistfully looking back to an idealized slower time and mourning its passing has long been the preserve of conservative political theory. Ironically, today, as William Connolly notes, the most virulent attempts to slow things down take the form of national and religious fundamentalism. Rather than rejecting modern speed, trying to turn the clock back, we should embrace the positive possibilities that speed contains for thought, judgment, human connection, and cosmopolitanism.[34] And, to do this, we need to direct our analytical gaze beyond the dialectics of speed to encompass the politics of technology itself.

Technology as Sociomaterial Practice

What role, then, does technology play in shaping people's experience of time? Does technological acceleration necessarily hasten the pace of our everyday lives? To answer these questions, let us look more generally at how the relationship between technology and society has been conceived.

The most influential commonsense assumption about the relationship between technology and society is "technological determinism." Few would explicitly subscribe to this theory, but, as I have indicated, it is pervasive. It has several versions, but in its strongest version, it is the claim that technological innovation is the most important cause of change in society. Key here is the idea that technology impinges on society from the outside, that technical change is autonomous and itself causes social change. That technology is not part of society but a separate, external sphere, so to speak.

By contrast, the founding principle of science and technology studies is that all technologies are inherently social in that they are designed, produced, used and governed by people. Perhaps it is worth saying at the outset that our objection to technological determinism was and is political as well as intellectual. Many of us who got involved in the development of this field in the 1980s had a simple polemical purpose, to shake the stranglehold that a naïve determinism had on the dominant understanding of the intertwining of society and technology. We were concerned that this view of technology, as an external force exerting an influence on society, narrows the possibilities for democratic engagement. It presents a limited set of options: uncritical embracing of technological change, defensive adaptation to it, or simple rejection of it. Against this, STS had its origins in a belief that the content and direction of technological innovation are amenable to sociological analysis and explanation, and to political intervention.

Being a critic of technological determinism does not entail a wholesale rejection of the profound influence that technological systems have had on the history of the twentieth century or on the way we live and who we are. No STS scholar would deny that technical innovation has social and cultural implications. Indeed, in *The Social Shaping of Technology*, we expressed some sympathy for a "soft" determinism: "to say that technology's social effects are complex and contingent is not to say that it has *no* social effects."[35] Rather, the aim has been to reconceptualize the relationship of technology and society. In doing so, we do not mean to understate the power of technology. Quite the contrary. Whatever version of STS we do, we do it because we passionately believe in the constitutive power of technology, that our societies and our very identities are shaped together with technologies.

For most social scientists, the recognition that technological change is profoundly shaped by social, economic, cultural, political, and organizational circumstances is too well established to need belaboring. The breakthrough contribution of STS scholars was to demonstrate that artifacts are socially shaped, not only in their usage but in their design and technical content.[36] Crucially, such an analysis rejects the notion that technology is solely the product of rational technical imperatives; that a particular technology will triumph because it is intrinsically the best. Technical reasons are vitally important. But we need to ask why a technical reason was found to be compelling, when it could have been challenged, and what counts as technical superiority in specific circumstances. Studies show that the generation and implementation of new technologies involve many choices between technical options. A range of social factors affect which of the technical options are selected, and these choices shape technologies and, thereby, their social implications. In this way, technology can be thought of as a sociotechnical product, patterned by the conditions of its creation and use.

In other words, technologies result from a series of specific decisions made by particular groups of people in particular places at particular times for their own purposes. As such, technologies bear the imprint of the people and social context in which they develop. It follows that political choices are embedded in the very design and selection of technology.

There is now a rich field of STS, reflecting a variety of approaches to the social study of technology.[37] I want to outline its distinctive perspective here, as I believe it provides the basis for understanding the complex intertwining of technology and time.

One way to begin is to think about things, the stuff we have, the material world rather than technology in general. Thinking about the use of things in this way connects us directly with the world we know rather than abstract ideas about the technology of our age. It encourages us to consider what we mean when we talk about technology and to be specific about *which* technologies have been most significant for acceleration.

For example, why do we immediately think about the latest digital gizmos when we think about speed? Why do the mundane material artifacts of everyday life, such as kitchen equipment, receive so much less attention in the narratives of technological progress? We live our lives surrounded by things, but we tend to think about only some of them as being technologies. It is common to think about technology as encompassing only very new, science-intensive things—ones with electronic or digital bits, for instance. And to think of the driving forces of history as being the steam engine, electricity, and computers rather than the washing machine, the stroller, and the condom. We tend to overrate the impact of new technologies in part because older technologies have become absorbed into the furniture of our lives, so as to be almost invisible.

Take the baby bottle. Here is a simple implement that has "transformed a fundamental human experience for vast numbers of infants and mothers, and been one of the most controversial exports of Western technology to underdeveloped countries—yet it finds no place in our histories of technology."[38] This technology might be thought of as a classic time-shifting device, as it enables mothers to exercise more control over the timing of feeding. It can also function to save time, as bottle feeding allows for someone else to substitute for the mother's time. Potentially, therefore, it has huge implications for the management of time in everyday life, yet it is entirely overlooked in discussions of high-speed society. This illustrates that the impact technologies are perceived to have on time largely depends on who is using them, and in what context, a point I return to below.

Social theories that announce the dawn of a new age are preoccupied with technology as major high-end technological systems rather than discrete tools or devices. The convergence of all kinds of electronic systems, broadcasting, telecommunications, and computer-mediated communications into a comprehensive ecology known as ICT is the classic case. But, even here, we need to consider specific applications in order to think about their effect on time.

Let us take the example of the Internet, perhaps the prime contender for a technology of acceleration. At its basic level, the Internet is indeed a set of technical standards and capabilities that enables a "network of networks" to interlink ICTs, including computers of all sizes and a variety of mobile consumer devices and electronic multimedia. The most visible aspects of the Internet are also technological: for instance, the tools that enable searches to be made through billions of pages of information on the World Wide Web (in a matter of minutes) or multiple e-mails to be sent to colleagues and friends.

Social research has shown, however, that the use of the Internet and its constituent and interlinking technologies form an intertwining, co-evolving web of people, social structures, and technologies. The paths opened by the Internet are determined not by technological capabilities alone but through a multitude of intricate social processes in which a diverse array of actors with varied goals participate in a rapidly evolving "ecology of games."[39] For example, the choices that are embedded in the operating codes of search engine software, such as Google, predispose users to access well-linked and highly connected websites and exclude poorly linked and less-connected websites.

This indicates that the Internet is about more than just equipment, and that the control of the Internet and related technologies is bound up with broader issues of who has access to the skills, equipment, and know-how essential to produce, use, consume, and govern the relevant technologies. Through this web of people and technology, the Internet can redistribute the relative "communicative power" of different actors in households, communities, workplaces, and society at large by reshaping access not only to information but also to people, services, and technologies.

Does it make sense, then, to think about the Internet as necessarily saving time? It certainly allows messages to be sent at amazing speeds. If technologies existed independently and outside of society, then surely faster technologies would save time as people would be doing the same things, but at a faster pace. At one level, this rings true—faster, more powerful computers enable us to process more information more quickly. Yet, paradoxically, we seem to end up with less time than before.

In reality, the impact of technological innovation is far from uniform or straightforward.[40] It is not only a matter of the presumed inherent capabilities of the technology in question. The extent to which its technical potency will be realized fundamentally depends on the social significance it is accorded and how it becomes embedded in its concrete and practical application. The Internet is especially open to manifold usages. In some instances, e-mail, for example, may genuinely encourage faster decision making, while in others, what is colloquially referred to as "information overload" may lead to inertia. Either way, what is clear is that technical velocity does not necessarily translate into more efficiency and convenience. As I detail in chapter 4, the significance of e-mail lies as much in the way it has transformed social expectations and standards of the time activities take as in the actual speed of communication.

My point is that qualities such as speed and efficiency are not produced by technologies alone but are related to social norms that *evolve* as devices are integrated into daily life. Rather than simply compressing time, information technologies change the very nature and meaning of tasks and work activities. Moreover, like the mobile phone, the Internet generates new kinds of material and cultural practices, reconfiguring the temporal and spatial basis of social interaction. It makes sense, then, to think about the relationship between technology and time as one of ongoing mutual shaping.

In order to understand this reciprocal process, I read technologies of all kinds as sociotechnical or sociomaterial "assemblages." This view, that technologies are combinations of people, materials, equipment, components, and institutions, is also sometimes referred to as an ensemble, a web, or a network. Whatever the term used, the idea is that the technical is not reducible to the social, nor is the social reducible to the technical. Whereas sociology tends to give primacy to social relations, treating them as existing prior to the intervention of technology, this perspective sees society and technology as mutually constitutive. In other words, the material world makes society possible.

Indeed, with the development of an STS approach known as *actor-network theory*, the agency of objects has taken center stage.[41] What does this mean? Although I have been emphasizing the key role of users in assigning meaning to objects, some formulations of this tend towards social determinism, that is, a view of technologies as infinitely flexible and tractable. Actor-network's conception of nonhuman objects as actors ("actants") or agents who exercise power serves as a corrective to this view. It helps us understand how the materiality and obduracy of artifacts create boundaries to the possibilities for interpretation and usage.

Consider for a moment the design phase of technological systems. In order to develop functional instruments or devices, engineers and designers anticipate the interests, skills, motives, and behavior of future users. Subsequently, these representations or configurations of users become built into or materialized in the form of the new product. In this sense, objects contain a "script" that stipulates how they can facilitate or constrain human activities, as well as relationships between people and things. This script delegates specific competencies, actions, and responsibilities both to users and artifacts. Bruno Latour's much-cited examples of automatic doors and road bumps illustrate how technical objects can impose programs of action on users, defining actors, the space in which they move, and the ways in which they behave and interact. Fittingly called *sleeping policemen*, road bumps are delegated the job of reducing motorists' speed where the rule of law does not suffice. In this way, it can be said that the material world itself exercises a kind of agency. Ironically, in this case, we see technology literally acting to achieve deceleration.

This is not to imply that the way artifacts evolve and are used is predetermined or irreversible. The linear model of technical innovation and diffusion, whereby innovation is an activity restricted to expert engineers and scientists, has long been revised. Long after artifacts leave the research and development laboratory, the process of design is still taking place. Users are no longer thought of as passive consumers of technology.

To take a well-known instance, the heavy use of mobile phones by adolescents exchanging SMS texts as well as audio messages was unanticipated by its designers. Or consider how developments such as Web 2.0, open-source software, and Twitter blur the traditional boundary between production and consumption. Innovative users can radically alter the meanings and deployment of technologies. While the technical character of devices matters a great deal, the life of machines ultimately depends on the locally contingent meanings that people attribute to them, in practice.

Conclusion

So to recap, a conception of technology as a sociomaterial practice recasts agency as emerging from the interconnections between people and things, the ensemble of human-machine interaction.[42] It moves the analytical focus away from examining how technology influences humans to exploring how materiality and sociality are constitutive of both activities and identities. From this perspective, people and things are not distinct entities but only exist in relation to each other. Technology and society are thus enacted together in a moving relational process achieved in daily "doings."

What implications does this have for understanding the dynamics of high-speed society? Does *technological acceleration* inexorably result in the *acceleration of the pace of life*?

My thesis is that time practices are always sociomaterial, that the contours and rhythms of our lives are calibrated by and with machines. In other words, we cannot comprehend the social organization of time separately from technology. Neither can we treat technology as a set of neutral tools with clearly defined, functional properties that determine unequivocally our time regimes. Instead, it is our human engagement with objects that generate those temporal qualities we tend to grasp as inherent in machines.

A social shaping perspective challenges the status and principal role accorded to technology in standard explanations of acceleration society. If technological change involves the interdependent activities of multiple heterogeneous agents, then it is necessarily a contingent process marked by contradictory consequences. The introduction of new machines always involves a dialectical process of promise, resistance, improvisation, and accommodation. Indeed, this process goes to the heart of who we are, as our very subjectivities and desires are articulated with technologies.

We experience the common feeling of being rushed as a personal condition that requires individual solutions. The natural reaction is to look to ever-faster technologies to save time. We are then mystified by the paradox of having more and better technologies and still feeling harried. It is easy to suppose that there is a temporal logic in digital technologies that is pushing us ever faster. Reifying technologies in this way makes us passive respondents to the instantaneous time of digitalization. Yet, as I will go on to demonstrate, the machines we have reflect our society and we are complicit in their design and implementation. Objects only take on their significance by way of our recurrent use of them. That is why there is no direct relationship between time saved and time gained.

Our newfound need for seamless connectivity, for example, can only be understood within the extended network of sociomaterial relations through which ICTs emerge and are stabilized as an ongoing practice. Digital technologies not only speed up information and communication but also open up wholly new domains of exchange, service, and entertainment. The range of options seems to increase in inverse proportion to our capacity to realize them. For Rosa, the acceleration cycle is self-propelling, as "the promise of acceleration never is fulfilled, for the very same techniques, methods, and inventions that allow for an accelerated realization of options simultaneously increase the number of options . . . at an exponential rate."[43] In my view, however, the way we choose to integrate such new activities and artifacts into our everyday lives depends upon the individual biographies and collective histories of both ourselves and machines.

Time is both intimate and social. It is the result of our collective entanglement with the material world. As such, it is infused with power relations, such as those of gender, class, and race, which are increasingly mediated by digital technologies. Our structural location and multiple identities provide us with differential technical resources and skills that alleviate or aggravate, increase or reduce, social distinctions and distances between us. Theories of the high-speed society mistakenly assume that acceleration is occurring across all sectors of society and all dimensions of life. But if disparate groups of people relate to both time and technology in diverse ways, then we need to explore how and why this is so. How we appropriate, adapt, and actively shape digital technologies to create new timescapes is an empirical question that I probe in the course of this book. By doing this, I hope to resolve the time-pressure paradox.

Perceptions of time do change with the emergence of new ideas and new inventions, but this always occurs in the context of preexisting ideas, habits, material apparatuses, and cultural practices. Understanding how time has been reshaped in the past may help us to become more alert to some of the profound changes in time consciousness that are still under way in our own times. As we will see in the following chapter, the idea of high-speed society is not as novel as we are led to believe.

GREAT NORTHERN Ry Co
No 441
MAKERS
1887
DONCASTER WORKS

Chapter Two

Time and Motion

Machines and the Making of Modernity

By failing to understand our own time system we also fail to fully understand ourselves.

HANNAH GAY, "Clock Synchrony, Time Distribution, and Electrical Timekeeping in Britain, 1880–1925"

We tend to think of high-speed society as a recent phenomenon associated with the growth of digital technologies. However, claims about technology annihilating time and space are not new. Here I will consider the impact of the vast technological changes that took place over the last century or so, some of which are arguably as sweeping as those wrought by the Internet.

My aim is not to dispute that contemporary temporalities have been radically altered with ICT, but rather to inform our understanding by adding a historical perspective. After all, arguments about the increasing pace of life only make sense against a putative slower past. We will see that our present-day embrace of speed has well-established antecedents. Indeed, much that is attributed to mass migration to the online world has roots sunk deeply in earlier phases of technical evolution.

In the modern world, timekeeping is an essential and habitual activity, and we constantly monitor and check the time. Social theorists trace our obsession with measuring time to the institutionalization of clock time under capitalism. The classic text is Edward Thompson's framing of clock time as disciplinary and as intimately related to the commodification of time as money.[1] The tyranny of clocks is essential to the narrative of the accelerating world. I will consider this argument in light of research that suggests that timekeeping has been a longer standing practice, and not necessarily a negative one.

While the time discipline involved in capitalist forms of production has played a key role in the shaping of modern time, by itself it cannot explain the changing cultural significance of speed. It is hard to exaggerate the effect of steam engines, railways, and the telegraph on people's experience of time-space compression. The railway journey, depicted by Charles Dickens and Marcel Proust, is iconic of a transformed sense of the tempo of everyday life. The traveler's view of landscape as a multitude of swiftly moving visual impressions was unprecedented.

However, it is only with the rapid shift to urban living that speed and change for its own sake have become celebrated as the general condition of modernity. Speed is one of several aspects of modern metropolitan life that becomes culturally valorized by artists and intellectuals, and it remains a key explanation of the time-pressure paradox. Georg Simmel's depiction of the emergent modern time consciousness as one involving immediacy, simultaneity, and presentism still resonates today. So too does his astute analysis of the alternate responses it would continue to incite, signaling both extraordinary opportunities and the corrosion of moral character. (Richard Sennett would later go on to identify short-termism as a unique consequence of new capitalism; see chapter 4) In my view, Simmel can well be reclaimed as the first theorist of the acceleration society.

In the twentieth century, the automobile came to symbolize speed, freedom, and liberty. It held out the possibility of change for groups and individuals who wished to escape traditional social confinement. The viability of boundless physical movement undoubtedly had political implications and examples of this will be discussed. However, as with all technology, the automobile's impact was far from straightforward. The same motorcar that promised unlimited movement also led to gridlock. And while the velocity of machines increasingly came to signify the driving force of progress and economic growth, the euphoria of constant motion also became associated with violence and destruction. These cultural contradictions remain central to the dynamics of acceleration.

Yet, the allure of pure speed continues to seduce. This is evident in the sociological turn to mobility, fluidity, and travel as the key descriptors of our present condition. Speed and perpetual motion are assumed to be universal desires of existence, realizable only through faster, more efficient machines. Accounts of this ilk are tied to linear narratives about the role of technical innovation in making modern times. In reality, technologies evolve through practical use and therefore come to mean quite different things to different people. Indeed, the ability of some to move fast and frequently can itself cause stasis for others. As such, technological acceleration is always accompanied by various kinds of slow down. It is no wonder, then, that our response to acceleration has always been characterized by profound ambivalence.

Punctuality and Progress

The need for people to coordinate their activities has been important throughout history, but never more so than today. We take for granted that our lives are shaped by the hours of the day, as measured by the clock. From childhood, we are taught the worth of punctuality, the imperative to be on time and not to squander it. The valorization of speed was central for the development of the industrial way of life. As Jeremy Rifkin remarks, "Efficiency and speed characterize the time values of the modern age. . . . The idea of saving and compressing time has been stamped into the psyche of Western civilization and now much of the world."[2]

Yet, what has come to seem natural and normal is the outcome of centuries of technical innovation and the circulation of ideas about time. Chronologies of standard technological history give the clock pride of place. And it is through the lens of Thompson's essay, "Time, Work-Discipline, and Industrial Capitalism," that social scientists primarily view modern time in economic terms, as market time.

According to Thompson, prior to industrialization, people depended on "natural rhythms" oriented to a variety of tasks related to an agricultural economy, task-oriented time. These older notations were replaced by a new manufacturers' time, a commodity measured in monetary terms and regarded as precious. He saw this as the result both of advances in timekeeping and of a Puritan ethic which helped people internalize the idea that time was not to be wasted. By the nineteenth century, according to Thompson, the idea of time thrift had become culturally embedded. Traditional rhythms began to look indolent and even primitive. The division and synchronization of labor became the norm in many manufacturing sites and, over the course of a century, manufacturers' time came to seem natural.

Time had become money. And, as Dickens foresaw in his novel *Hard Times*, while clock time standardizes time, not everyone's time is of equal worth. As Mr. Bounderby, the businessman, put it to the freedom-loving circus people, "We are the kind of people who know the value of time, and you are the kind of people who don't know the value of time." To which the circus performer, Mr. Childers, retorted, "If you mean that you can make more money of your time than I can of mine, I should judge by your appearance, that you are about right."[3]

Throughout their extensive publications, both Barbara Adam and Helga Nowotny trace the historical shift in the way people understand time to the clock culture that developed in modernity. They stress that industrial time engendered the pursuit of a disciplined and frugal use of time in the quest for efficiency. Time became commodified, compressed, colonized, and controlled.[4] And once the linear system of time was set:

> Acceleration could start in the form of motion making everything dynamic, which seemed to stop at nothing. In the *tourbillon social* which broke out with the industrial revolution and wrenched people out of their countless "small worlds" . . . acceleration became the experience of modernisation overshadowing and shaping everything else. The pace became more important than the destination: anyone who stands firm stands still; everything, above all time, becomes frantic motion: the new myth was speed.[5]

Marx's analysis of the commodification of time remains the touchstone for such writing.[6] Marx's central argument was that an empty, abstract, quantifiable time, applicable anywhere, anytime, was a precondition for its use as an abstract exchange value on the one hand and for the commodification of labor and nature on the other. Only on the basis of this neutral measure could time assume such a pivotal position in all economic exchange. Since "moments" are "the elements of profit," it is command over the labor time of others that gives capitalists the initial power to appropriate profit as their own.[7]

Struggles between owners of labor and capital over the use of time and the intensity of labor have been endemic. What Thompson's account highlights is that it took several generations for the new labor habits and the new time discipline, "the familiar landscape of disciplined industrial capitalism, with the time-sheet, the time-keeper, the informers and the fines," to be instilled. The time discipline was based on obedience to the clock and to the appointments specified on it, such as the time to begin work. In a much quoted passage, Thompson records how two nineteenth-century factory workers testified that they were not allowed to have their own clocks or watches on company grounds:

> In reality there were no regular hours: masters and managers did with us as they liked. The clocks at the factories were often put forward in the morning and back at night, and instead of being instruments for the measurement of time, they were used as cloaks for cheatery and oppression. Though this was known amongst the hands, all were afraid to speak, and a workman then was afraid to carry a watch, as it was no uncommon event to dismiss any one who presumed to know too much about the science of horology [clock and watch making].[8]

Furthermore, if people could be taught the new time discipline early in life, they would be better prepared to meet the growing synchronization demands of the workplace. As a result, there was an increased emphasis on teaching punctuality in schools in both England and the United States. Such practices continued in the twentieth century. Allen Bluedorn notes that in 1903, his American maternal grandmother, at the age of thirteen, received a school attendance certificate with the telling phrase "having been neither absent nor tardy during the month ending."[9] To this day, meeting attendance requirements is an essential criterion for graduation at the London School of Economics.

In recent years there has been a lively discussion about the historical accuracy of the major prevailing accounts of clock time. In their volume *Shaping the Day: A History of Timekeeping in England and Wales 1300–1800*, Paul Glennie and Nigel Thrift critique the technological determinist version based on drawing direct relationships between technical developments in clocks and the hegemony of clock time.[10] They also maintain that Thompson's account draws too direct a relationship between economic changes and clock time. Echoing my own perspective, the authors argue that while we need to take the devices seriously, it is important to understand clock times as everyday *practices*, which were (and are) remarkably diverse.

There is mounting evidence of the widespread use of clocks and timekeeping practices from the fifteenth century onward, long before private ownership of clocks and watches was commonplace, let alone the emergence of factories. In early modern England, for example, diaries and letters indicate that schools were already imposing a temporal discipline: "Now at five of the clock by the moonlight I must go to my book and let sleep and sloth alone" ran a saying that dates from around 1500 and is attributed to a twelve-year-old boy.[11] Whether he actually arrived on time is lost to history. Nonetheless, there was a timetable and an intended discipline, known to the young, confirming "the sheer density of temporal infrastructure" at that time.

The issue of periodization need not detain us, but the case that Glennie and Thrift make is salient to my argument here. They eschew a linear view whereby early forms of timekeeping were wholly replaced by clock time. Instead, they stress the very different registers and dimensions of overlaying clock times that coexisted in the past, as they were embodied in complex sets of practices in different temporal communities. There was no sudden rupture, whether for better or worse, with the advent of industrialization. We were and are able to internalize and live with many different time notations, astronomical, biological, private and public, and so on.

It follows, therefore, that we cannot accept the influential commentaries about the preeminence and triumph of clock time. The authors subscribe neither to a story of technological supremacy in the later eighteenth century, heralding a glorious modernity, nor to a jeremiad on the imposition of strict timetables and the loss of preindustrial freedom.

Indeed, they firmly reject the increasingly popular view that the procedures and practices of aggregation that we call "clock time" are to blame for all the ills of the world: "clock time has been as much a liberatory as an oppressive force. It has allowed as much as it has disciplined. New entities, capacities, and experiences have become possible which did not exist before and there is no reason to believe that all of these have been negative."[12] Fear of the omnipotence of clock time runs through contemporary narratives about the accelerating world and betrays a tragic teleological tale about modernization.

The continuing role of clock time in relation to labor productivity is a theme I will pursue in chapter 4. There is certainly much evidence that the use of clocks to increase the rate of work was even more marked in the early years of the twentieth century, with developments such as scientific management and Taylorism.[13] It would later be immortalized in Charlie Chaplin's 1936 film *Modern Times*, which depicts the alienated factory worker literally fighting the clock to slow down production.

What is undeniable is that by the early twentieth century a new sense of temporal exactitude was developing. Inhabitants of that era would have been conscious of a change in the pace of working and social life; of living in a culture in which accurate timekeeping, efficiency and punctuality were becoming normative.

However, in order to understand how speed enters the modern cultural imagination, we need to move beyond a narrow focus on clocks as talismanic artifacts. Arguably, communication systems like the railway, the telegraph, the telephone, and wireless communications were as significant as increasingly accurate timepieces in leading to novel ways of experiencing both time and human affairs. As we shall see, this new material world paradoxically brought both an optimistic sense of security, of being in control of events, and a sense of insecurity, a sense of a world speeding out of control. And, as Hannah Gay observes, "In the early twentieth century these two sensibilities ran in parallel and played off each other."[14]

Machine Speed and Modernity

Much has been written about the massive social and technological changes that took place from the mid-nineteenth to the mid-twentieth century. The cultural historian Stephen Kern's well-known account describes how innovations in art, architecture, literature, science, and technology mutually influenced and inspired each other.[15] The achievement of higher speed was integral to those developments and permeated every aspect of society. Speed, in the sense of modern machine velocity, formed a powerful cultural narrative during this period, yoking together machines, money, and progress. The telegraph presaged many of these changes, so altering the structure of social relations that it has been christened the Victorian Internet.[16] I will therefore consider this case before turning to look at modernist avant-gardists' celebration of speed.[17]

But first it is befitting to reflect on the abiding connection between machine speed and progress. Although the idea of progress can be traced back to the classical world, it only becomes a powerful social ideology in the first part of the twentieth century. Our common sense notion of "modern" denotes a historical process of steady advance and improvement in human material well-being, occasioned by technological innovation. How, though, asks John Tomlinson, does speed itself become a prime mark of *social* progress?[18]

One answer is the straightforward association between the pace of mechanical production and the delivery of material improvements. The speed of manufacturing, transportation, and communication technologies saved vast amounts of physical effort and time as well as providing affordable material goods. For the first time, human ingenuity deploying mechanical power appeared to overcome the natural order, giving rise to engineering notions of control and regulation. Rational mechanical speed promised to overcome the physical realities of space, distance, and separation—obstacles to the fulfillment of human needs and desires.

In this way, speed presents itself as the prime condition for economic growth and prosperity. The associated increase in the pace of life, though it may not be attractive in itself, may appear as a matter of "pragmatic acceptance as part of the cultural 'bargain' with modernity." However, there is also a quasi-moral linkage between speed as dynamism and visions of the human good. The ideological nub of progress, Tomlinson argues, is its *impatience* with the way things are, that human good lies in the struggle for improvement. Change thus comes to be valorized over continuity, and once this is accepted, the speed of change becomes a self-evident good. "This moral underpinning of mechanical speed combines with the material benefits if offers and its sheer excitement, to construct a hugely powerful cultural narrative of social acceleration."[19] That notions of speed and progress are still so intertwined in contemporary political discourse is integral to the insistent sense of time pressure.

Although our own experience of time-space compression is unique in detail, its structure is characteristically modern. The dramatic effect of the electrical telegraph on the mental maps of Europeans and Americans is illustrative of this. For the first time, a communication machine could separate communication from transportation, allowing information to move independently of—and much faster than—transport. It caused people to wonder, much as the Internet does today, about the rapid and extraordinary shifts it wrought in the spatial and temporal boundaries of human relationships. Indeed, James Carey argues that "the innovation of the telegraph can stand metaphorically for all the innovations that ushered in the modern phase of history and determined, even to this day, the major lines of development of American communications."[20]

Compared to the telegraph, the Internet does provide a spectacularly enhanced degree of speed. However, many of the claims for the revolutionary consequences of the Internet presume a putatively different past, and a belief that our current ambivalence toward technological change has no precedent. For example, the emergence of global space is not as new as we think. The telegraph also promised to annihilate time and space and to bind all of mankind together "on the face of the globe." In the words of the British prime minister, Lord Salisbury, in 1899, the telegraph "has, as it were, assembled all mankind upon one great plane, where they can see everything that is done and hear everything that is said, and judge of every policy that is pursued at the very moment those events take place."[21] Likewise, the consequences of this for the very nature of language, knowledge, and human awareness led to both the kind of euphoric claims and accusations of trivialization that bear an uncanny resemblance to discussions about Twitter.

The telegraph had a profound impact on the conduct of commerce, government, the military, and colonialism, dramatically altering the ways in which time and space were understood by ordinary men and women. The telegraph's role in establishing standard "railway" time is well known. The eventual adoption of Greenwich Mean Time brought the world within one grid of time, uprooting older, local ways of marking the passage of time. Similarly, standard units of distance and territorial measurement incorporated space into a single regime of measurement.

It is precisely these transformations that Anthony Giddens famously situates at the heart of the constitution of modernity. The new systems of calibration provide the means for "precise temporal and spatial zoning" and thereby produce new topographies and chronologies of experience, new divisions between public and private space, work and home, labor and leisure, employment and retirement.[22] In other words, the very dynamism of modernity "derives from the separation of time and space and their recombination." The result is a fundamentally changed consciousness of temporality in social and cultural life.

However, Giddens is unclear about the precise forces that produce these shifts. His discussion moves between offering a metatheory of space and time and a focus on how modern societies actually organize these dimensions. Either way, he does not pay much attention to the role of communication systems in shaping the modern experience.[23] This is at least in part because he treats technology as an autonomous force rather than as a sociomaterial ensemble of humans, machines, infrastructures, institutions, and everyday practices. The modern experience of time was actively reconstituted together with the technologies that fostered it. And the commodification of time that underpinned industrial capitalism relied on a whole range of interconnected technological innovations.

It is salutary, then, to recall the less familiar tale of the telegraph's role in the development of commodity markets. According to Carey, the telegraph was the critical instrument in making time the new frontier for commerce. Before the telegraph, markets were relatively independent of one another and the principal method of trading was arbitrage: buying cheap and selling dear by physically moving goods around. When the prices of commodities were equalized in space as a result of the telegraph, however, commodity trading moved from trading between places to trading between times, shifting speculation from space to time, from arbitrage to futures.

In eliminating space as an arena of arbitrage, therefore, the growth of communications gave rise to the futures market. In order to develop, futures markets required three conditions: that information moved faster than products, that prices were uniform in space and decontextualized, and that commodities be separated from the receipts that represent them and be reduced to uniform grades. The shift of market activity from certain space to uncertain time was, Carey remarks, "the first practical attempt to make time a new frontier, a newly defined zone of uncertainty, and to penetrate it with the price system."[24] In a sense, the telegraph invented the future.

The Allure of Speed

No wonder, then, that central to the intellectual projects of many European thinkers of the early twentieth century was a radical questioning of the Newtonian world of calculable, linear time, and space. From H. G. Wells's classic novel *The Time Machine* (1895) to Albert Einstein's papers (1905), many of the questions of the new century were centered around the malleability of space and time.[25] Could time be stretched or compressed? Could it be accelerated or reversed? Was time perceived differently by different observers, and, if so, could there be a universal time?

Most commentators agree that there was a maelstrom of creativity in the aesthetic realm during this period. A whole new world of representation and knowledge resulted, which qualitatively transformed what modernism was about. Inventions such as the telephone, wireless telegraph, X-ray, cinema, the automobile, and the airplane led to major material changes in daily life and precipitated new modes of thinking about and experiencing space and time. David Harvey persuasively argues that the simultaneity derived from this rapidly changing experience had much to do with the birth of modernism. And to be modern, Marshal Berman reminds us, is to "find ourselves in an environment that promises us adventure, power, joy, growth, transformation of ourselves and the world—and, at the same time, that threatens to destroy everything we have, everything we know, everything we are."[26]

It is in relation to speed that I want to examine this distinctly modern experience of living in and with profound ambivalence. Where better to start than with the Futurist Manifesto's (1909) declaration that: "the splendor of the world has been enriched by a new beauty: the beauty of speed. A racing automobile with its bonnet adorned with great tubes like serpents with explosive breath . . . a roaring motor car which seems to run on machine-gun fire, is more beautiful than the Victory of Samothrace"? This was one of the first documents to celebrate the automobile as an object of beauty and to cite speed and acceleration as aesthetic elements.

Futurism was an artistic, cultural and social movement that passionately embraced the future, exalting speed, power, technology, youth, and violence. As its founder, Filippo Tommaso Marinetti, continues, "Time and Space died yesterday. We are already living in the absolute, since we have already created eternal, omnipresent speed." The movement was part of the modernist avant-garde of the early twentieth century whose followers sought to revolutionize everyday life by leading by example. Although they issued party-like manifestos, they did not usually seek change by violent means. The Italian futurists' protofascist glorification of war was the exception, while the Russian avant-gardists tended to act in support of the Bolshevik Revolution until the suppression of artistic autonomy in the Soviet Union.

A common feature of the avant-gardists was their wholesale rejection of the past, of everything old, and an exuberance about "the new." Accordingly, they were especially enthusiastic about new technologies, such as the car, the airplane, and the industrial city. Steel, concrete, and sheet glass were preferred over brickwork, the legacy of which can be seen in every major city of the world. From the outset, modernist architects sought to rebuild the urban landscape through rational planning and engineering in order to deliver an enhanced, dynamic lifestyle. As Le Corbusier famously said, "A city made for speed is made for success," and the car was therefore integral to its design.[27]

Italy was the birthplace of futurism, and it was here that the first motorways were built. By the end of the 1920s they covered over four thousand kilometers and were touted by Benito Mussolini as one of his greatest achievements and proof of his commitment to progress and modernization. The pure hedonism of speeding on a motorway would later be captured in the electronic, repetitive rhythms of the German technoband Kraftwerk's song "Autobahn," named after the expressway system, which uniquely has few speed limits.

In extolling the virtues of rational, functional planning, what modernists like Le Corbusier overlooked was the fundamental ambiguity that the urban experience would induce. It was the German sociologist Georg Simmel who anticipated the nervous stimulation and sensory overload generated by the tempo of metropolitan life, as we shall see. But first, I want to pause for a moment to describe the two opposing facets of machine speed itself—that it represents both economic growth and violent destruction.

The futurists best captured the aesthetic excitement of those transformations in everyday life associated with the new celebratory culture of machine speed. In doing so, they identified three core elements of the modern twentieth century's cultural imagination, as Tomlinson outlines:

1. That the sensual-aesthetic experience to be derived from fast machines is valuable and desirable in itself and that the risk and danger associated with speed offers satisfactions beyond those generally sanctioned within mainstream society.
2. That courting this risk and danger has an "existential"/heroic/transgressive dimension.
3. That speed and violence are inextricably intertwined.[28]

Why people find speed itself intoxicating is a complex psychological matter. My interest is rather in the cultural association between machine speed and sensuality, risk, thrills, and danger. I have already mentioned the essential affinity between speed and modern warfare in relation to Virilio. The more general point about the emotional power of "affiliative" or "evocative" objects, and the pleasures that can be derived from the mastery of machinery, is a long-standing theme in science and technology studies.[29] My own technofeminist writing has examined the gendered nature of this technical culture, not something Tomlinson touches on, and I will return to it later.

Tomlinson's observations about the contradictory impulses of capitalist modernity, that "the impulse to promote speed in one area of life begets the need to regulate, even to suppress it, in others" are, however, astute.[30] The resulting tensions of life in an accelerated culture are vividly illustrated by driving. On the one hand, there is the powerful mythology of iconic rebels such as James Dean, who live recklessly and die young at the wheel, and heroic individuals like Chuck Yeager, the first pilot to travel faster than sound (revered in Tom Wolfe's novel *The Right Stuff*). On the other hand, driving has become a mundane, everyday necessity, and, in a time-pressured culture, people's desire for speed is incessantly frustrated by speed limits and traffic congestion that inhibit them from driving fast.

But speed is literally lethal. The car is an instrument of violence and destruction, a vehicle of "mass murder," according to Norbert Elias.[31] The World Health Organization estimates that well over one million people are killed on the road every year. This varies for different countries, with high rates in Latin America and Africa. Even so, road safety is largely seen as the sole responsibility of individual road users. Automobility only "works" because its accidents are denied.[32] The traffic accident is not seen as a normal social occurrence, but as an aberration. While there is a morbid fascination in the media with spectacular car crashes, and they are the stock in trade of Hollywood action movies, their routine occurrence is only news to the extent that they disturb traffic flow.

Car accidents are, however, predictable and preventable. As any STS scholar would tell you, the technical solution to dangerous driving is not a speed camera or Latour's iconic road bumps. The way to regulate speed is to design slower cars. But car engines are specifically designed and marketed for their capacity for high velocity and fast acceleration. Cars are not only transport machines but also intimate objects expressive of individuality and lifestyle choices. (Even I must admit to being seduced by the allure of speed, having owned an MG sports car as an eighteen-year-old in Australia and enjoying its close-to-the road feel.) According to J. G. Ballard's novel *Crash*, the car crash may even be a source of sexual fetishism. The incongruity of the automobile's promise of freedom of movement with the actuality of a largely sedentary existence in a landscape dominated by traffic-overloaded motorways is even more pronounced today. However, it is the speed of information flows rather than of motorcars that is at the forefront of our imagination.

The Metropolitan Pace of Life

A sense of acceleration has thus accompanied the path of Western modernity since its origins. The modern metropolis is the prime site for the intensification of time use, as it creates a dense set of possible interactions in a small space. As we move into an era where more than half the world's population lives in cities and the number of global cities has mushroomed—estimated to number about seventy worldwide—the urban experience is becoming ever more pervasive.[33]

It is in this context that Simmel's writings have once again become resonant. For Simmel, in contrast to other social theorists, it is the metropolis rather than the industrial enterprise or production or rational organization that is key to modernity. His insights about the increasing pace of life in *fin de siècle* Europe, his take on the zeitgeist, have such affinity with postmodern discussions of our contemporary condition that I want to recall them here.

In *The Philosophy of Money*, Simmel analyzes the ephemerality and briefness that have become signifiers of the temporality of modernity. For him, there is an intrinsic connection between the increased pace of life in the city and the peculiarity of money. Indeed, he draws a direct parallel between the effects of the mathematical character of money and the general use of pocket watches: "like the determination of abstract value by money, the determination of abstract time by clocks provides a system for the most detailed and definite arrangements and measurements that imparts an otherwise unattainable transparency and calculability to the contents of life, at least as regards their practical management."[34]

Money only fulfills its function through its circulation, and thus it speeds up every activity connected with money, making them continuous. Production, transportation, sales, or consumption all have to be constantly on the move, and this revolutionizes the time-space coordinates of social relations. The totally dynamic impetus of the money economy, throwing everything into the circulation process, shatters stable and constant relations and creates a transitory constellation of relations in which everything is in flux, with no secure resting points.

The perfect institutional embodiment of the "teleology" of money, as an "end in itself," is the stock and commodity exchange, where time is radically compressed, and "values," in Simmel's words, are "rushed through the greatest number of hands in the shortest possible time." The human activity of the exchange is emblematic of the larger social trend, namely "an extreme acceleration in the pace of life, a feverish commotion and compression of its fluctuations, in which the specific influence of money upon the course of psychological life becomes most clearly discernible."[35]

What is particularly fascinating is Simmel's description of the modern personality types that this social turbulence generates. The classic metropolitan type, the blasé individual, suffers from the "*intensification of nervous stimulation* which results from the swift and uninterrupted change of outer and inner stimuli."[36] The cornucopia of possibilities and distractions available within the capitalist metropolitan milieu make it the locale in which "stimulations, interests, fillings in of time, and consciousness" are offered in profusion. In stark contrast to the slow rhythm of rural life, each crossing of the city street creates "the rapid crowding of changing images, the sharp discontinuity in the grasp of a single glance, and the unexpectedness of onrushing impressions." While too much nervous stimulation can cause the blasé character to become obsessive and even pathological, for Simmel, these same processes are also responsible for "the finest and highest elements of our culture." Indeed, his critical perspective did not prevent him from fully appreciating the myriad stimulations of the vibrant urban scene and the enlarged social horizons, freed from tradition, it proffered.

The cultural value we attach to having a busy lifestyle, one rich in manifold activities and events, echoes this sentiment and it is a recurring theme in this book. As one of the leading time-use scholars, Manfred Garhammer, argues, the "*ambivalent* consequences of modernity distinguished by Simmel are crucial for the understanding of the time-crunch-life-enjoyment-paradox: life may become richer in terms of the number of events, and at the same time it may become poorer."[37]

Simmel was acutely aware of the inherent ambiguity of modern city life, that it promotes individualization and standardization at the same time. For example, in his exploration of fashion and style, we find the dialectical interplay of individual imitation and differentiation, the desire to be like others and the desire for difference. Fashion requires continuous reproduction to accelerate the turnover time of new commodities, making its newness its simultaneous death. As such, it exemplifies the modern cultural fixation with the "eternal present," with immediacy, the transitory, perpetual motion. Simmel was thus highly attuned to the emergent time consciousness of the modern individual: "the dominion of presentism—erasure of the past, effacement of inherited connections, domination by the immediately invisible sublime—is an integral part of modernity."[38] The point I will return to is that this condition of immediacy is taken today as being wholly a consequence of digital technologies.

Strikingly absent from Simmel's prescient analysis, however, is how the experience of the metropolis is highly stratified by social class, status, gender, and ethnicity, in other words, power relations. The realization of simultaneity as a phenomenon of the perception of time, the awareness of everything happening in the moment, was reserved for the privileged few. The acceleration of the pace of life was not then, and is not now, a uniform condition of existence.

Speed and mobility remain differentially distributed, accessed and interpreted by different groups depending on their circumstances. If we return to the subject of the automobile, we will see that this technology itself shifts from being a sociomaterial practice of the rich to becoming, literally, a vehicle of democracy in the era of mass consumption. Once again, the speed it offers has unanticipated consequences.

Automobiles: Wheels in Motion

The automobile is a preeminent feature of the urban environment. The modern city is premised on car travel, and it was the mass production of the motorcar that greatly influenced its shape. Wide access to the experience of automobile speed was, in turn, enabled by the acceleration of car production.

Henry Ford did not invent the motorcar. Nor was his Model T a particularly good motorcar. Ford was not even the first to use a moving assembly line. But he was the first to "mass produce" a car, a phrase that he was also the first to use. As a result, the time taken to assemble a Ford chassis fell from just under 12.5 hours in the spring of 1913 to 93 minutes a year later. Greater efficiency led to big falls in price: the Model T cost $950 in 1909 and $360 in 1916, a fall in real terms of more than two-thirds.

Ford realized his aim of building a car "so low in price that no man making a good salary will be unable to own one." Between 1908 and 1927, Ford sold a total of 15 million Model Ts. Other car manufacturers followed suit, so that between 1908 and 1923 the average price of a car fell from $2,126 to $317 (in 1908 dollars). At the same time, annual sales rose from just 64,000 to 3.6 million. For its period, this rate of diffusion was extraordinary. Indeed, economic historians Tim Leunig and Hans-Joachim Voth argue that mechanizing the production process of the car (as with cotton spinning) was as valuable in terms of consumer welfare as inventing the Internet, and much more valuable than inventing the mobile phone.[39]

Weighing up the costs and benefits of the motor vehicle is not easy. Its significance and the lived experience of driving are full of complexity, ambiguity, and contradiction. Undoubtedly, in the early twentieth century, the car represented freedom for many and, arguably, had a greater impact on women than on men. Notably, the automobile appeared at the same time that women were striving for freedom in the home and in politics. At first, women were almost exclusively passengers. It was the electric automobile that gave upper-middle-class women the freedom to leave home and break free of the control of their husbands. In her book on women and the car, Virginia Scharff traces the critical role of the automobile in facilitating the suffragettes' mobilization, allowing the regional and cross-country campaigns that led to women winning the right to vote in the 1920s.[40] However, women had no place then in the actual manufacture of cars, while from early on the racial division of labor was built into the Ford factory.

The liberation and autonomy promoted by the private car also played an important role in the civil rights struggle, as Paul Gilroy describes. Initially, automobiles had been exclusively presented to white consumers and some companies expressly stipulated that their machines not be sold to blacks. However, when they could afford to buy them, African Americans bought cars at least as readily as their economic circumstances permitted. At one level, they were an absolute necessity for finding and maintaining employment. But the car also acquired additional significance in that "for African American populations seeking ways out of the lingering shadows of slavery, owning and using automobiles supplied one significant means to measure the distance traveled towards political freedom and public respect."[41] No wonder that feelings of rapture and kinesthetic pleasure, of being in control of so much power and speed, would feature so strongly in black music and culture. That the car was frequently linked to the female body and driving to sex is another aspect of its gendered politics.

The popular promise of automobile speed was to be short lived as more and more cars hit the road. One billion cars were manufactured over the course of the twentieth century, and there are currently over seven hundred million cars moving around the world's roads.[42] The automobile and its infrastructure dominate most North American cities in the literal sense that vast tracts of land are required to accommodate it. Not only for the roads, but also for bridges, service stations, and parking spaces—at home, work, the supermarket, and everywhere that people congregate. Small wonder that in American cities, close to half of all urban space is dedicated to the automobile; in Los Angeles, the figure reaches two-thirds.

For the individual, the mobility and convenience that the private car bestows are unparalleled by any other means of transportation. However, what appears to be an ideal solution to individual needs is increasingly illusory as more and more people choose, or are forced to make, similar decisions. In terms of individual mobility, the utility of the motor vehicle is diminishing as the number of cars on the road escalates. The prosperous 1950s and early 1960s were characterized by booming car ownership and, at least in the United States, the car was expected to be the future of urban transport. The land use/transport planning procedures of that period pioneered the building of elaborate highway and freeway systems. But freeways themselves spawned more and more traffic, until very soon after their completion they were already badly congested. The obvious response to traffic congestion was to build more roads, leading to a vicious cycle of congestion, road building, sprawl, congestion, and more road building. The drive to save time proved somewhat counterproductive.

The net result is that London rush-hour traffic averages about ten miles per hour; in Tokyo, cars average twelve miles, and in Paris, seventeen. Indeed, riding a Victorian technology in central London—the bicycle—during peak hours is faster than traveling by car.[43] By comparison, the average daily travel speed of thirty-three miles per hour in Southern California, where there are probably more miles of freeway than anywhere else in the world, may seem impressive. However, as a result of a much lower population density than European cities, the advantage of speed is offset by the much longer distances required to travel to work.

The irony is that a horse and buggy could cross downtown Los Angeles or London almost as fast in 1900 as an automobile can make this trip at 5 p.m. today. Similarly, the speed of air transport has a wavering history. Whereas by 1958 travelers were going five hundred and fifty miles per hour in a Boeing 707, today we go no faster than in 1958 but rather slower due to the need to conserve fuel.[44]

I noted above how destructive car speed is in terms of road deaths worldwide. It is also a major cause of environmental pollution and international conflict in the imperial pursuit of fossil fuel. What is less remarked on are the ways in which the dominance of automobiles can lead to slow-down. The incongruity of the private car is that the accelerated mobility achieved in the "conditioned atmosphere and internal music of this windowed shell" is predicated on the sedentary human body.[45] Rather like the experience of surfing the Internet at a computer screen, the driver is both stationary and mobile at the same time.

In affluent societies where the automobile dominates, the car is deeply entrenched in the ways in which we inhabit the physical world. In redefining movement, the car contains contradictions, as Virilio warns us, speed, gridlock, and a sedentary lifestyle. Indeed, the car is profoundly implicated in the recent finding that much of the world's population is physically inactive. According to the medical journal *Lancet*, this increases the risk of many adverse health conditions, such as cancer and diabetes, and is therefore a major public health issue.[46] Worldwide, it is estimated that physical inactivity causes 9 percent of premature mortality, or more than 5.3 million of the 57 million deaths that occurred worldwide.

There are many other dimensions to the unequal access to speed and movement. The postcolonial metropolis is host to a massive discrepancy in material conditions and life chances, and for most of the world's population, their only experience of speed is on a bicycle. In fact, nowadays many more bicycles than cars are being made, largely due to the massive expansion in Chinese production. The migration of millions from the countryside in places like China is a major source of global population movement and it is a far cry from the pleasures of the urban explorer evoked by the iconic modernist figure of the *flâneur*.

Conclusion

I would like to end this chapter with some reflections on the upsurge of academic interest in the significance of flows, movement and mobility in social life. John Urry, for example, has called for a new sociology of mobility, that sociology should focus on movement, travel, and mobility as opposed to settled bounded institutions.[47] He argues that fundamental global transformations are making the concept of society less analytically useful. We are better off analyzing the social in terms of flows and networks, as mobility is now the determining feature that frames social relations, not structures or positions. This emphasis on mobility is also a key feature of Zygmunt Bauman's writings about liquid modernity.[48] There is much talk of the ubiquity of various forms of travel, that the paradigmatic modern experience is that of rapid mobility over long distances, while migration is often represented as the central global phenomenon.

However, this model of contemporary life is, in fact, strictly applicable only to a relatively small number of highly privileged people. As David

Morley notes, "Despite all the talk of global flows, fluidity, hybridity and mobility, it is worth observing that, in the UK at least, there is evidence that points to continued geographical sedentarism on the part of the majority of the population."[49] Over half of British adults live within five miles of where they were born. Even in the more geographically mobile United States, two out of three people do not have a passport.

We should therefore not exaggerate the role of long-distance travel in people's lives. Despite globalization, local life occupies the majority of our time and space, and vast realms of people remain static, whether through choice or force of circumstance. Some groups are more mobile than others and have more control over both their own mobility and that of others. The mobility available to the affluent middle classes is quite different from the mobility of the international refugee or migrant, domestic worker.

Speed for the few is contingent on others remaining stationary. As Tim Cresswell remarks, "Being able to get somewhere quickly is increasingly associated with exclusivity."[50] Even in air travel—where all classes of passenger travel at the same speed—those in first class pass smoothly through the airport to the car that has been parked in a special slot close to the terminal. London's City Airport offers business travelers to New York the option of flying via Dublin for immigration clearance in order to be in the fast lane on landing. Meanwhile, the majority of foreign arrivals are left waiting in the slow lanes. "Speed and slowness are often logically and operationally related in this way."

Theorists of mobility flatten out such differences because they mistake their own partial experience for a universal condition. The same can be said of much of the literature on social acceleration. The effect is to legitimate, Bev Skeggs argues, "the habitus of the middle-class that does not want to name itself."[51] Voluntary mobility, like speed, is seen as a social good, while fixity becomes associated with failure, with being left behind. Moreover, the notion of mobility is itself gendered in various ways such that women occupy a fixed place in the male narratives of travel, adventure, and discovery. Just as the literature on modernity, describing the fleeting, anonymous, ephemeral encounters of the life in the metropolis mainly accounts for the experience of men, so these arguments ignore the lingering separation of public and private spheres.[52] By equating the modern with the public, they fail to describe women's experience of immobility. As with time, mobility is a resource to which not everyone has an equal relationship.

Summing up, the idea that the pace of life is accelerating is not new. The vast technological and social transformations that took place in the early twentieth century were also accompanied by the experience of time-space compression. These sociomaterial changes in the fabric of everyday life were to be consolidated in the modern city. Life lived at a high speed became identified with progress. Our valorization of a busy lifestyle, as well as our profound ambivalence toward it, can be traced back to this period.

Technology played a key role then as now. The introduction of the automobile in particular symbolized machine speed in the twentieth century and, like the telegraph and telephone, it reduced barriers of distance and made the world more interconnected. However, its promise of pure speed proved somewhat hollow. In any event, constant movement as a shared aspiration of the good life hides many realities. Speed and mobility are far from a universal condition of existence.

Nevertheless, the experience of immediacy as a phenomenon of the perception of time, the awareness of everything happening in the moment, has become widespread. Whereas a hundred years ago it was the preserve of the privileged, almost everyone has become implicated, at least potentially. Worldwide simultaneity is now the taken-for-granted condition of our lives and is what the Internet lays claim to.[53] The social shaping of technology and time is constantly evolving, and we need to specify how these processes coalesce in order to gain a more balanced understanding of our current digital times.

My intention here has been to show that an awareness of what our social, economic, and technological arrangements owe to the past makes the acceleration society thesis more intelligible. Connecting with these earlier debates brings the realization that the questions we face today are not in themselves new. This does not detract from their urgency. But in order to develop a critical perspective on the discourses of acceleration that surround us, we need to put them in a fuller historical perspective than that which is generally acknowledged.

Throughout this chapter I have argued that the discourse of acceleration tends to skim over and conceal the extent to which the pace of modern life depends on one's resources and the choices they make possible.

In reality, both control over time and access to mobility reflect and reinforce power. Skeggs is rightly critical of the universalizing treatise of the mobile, cosmopolitan individual postulated by social theorists, such as Ulrich Beck and Anthony Giddens, and she notes the ever-widening gap between those who theorize and those who engage in empirical research.[54] It is only by examining the facts on the ground that we can understand how speed and time are being lived. So this is exactly what we will do in the remaining chapters of this book.

Chapter Three

The Time-Pressure Paradox

The time we have to spend each day is elastic: it is stretched by the passions we feel; it is shrunk by those we inspire; and all of it is filled by habit.

MARCEL PROUST, *In Search of Lost Time*

The ability to choose how you allocate your time lies at the core of a positive notion of freedom. Idleness and abundant leisure were once markers of the aristocracy. Today a busy, frenetic existence in which both work and leisure are crowded with multiple activities denotes high status. However, just as in the past, people's control over their own time largely depends on their personal circumstances and financial resources. While this is equally true of our relationship to technology, this chapter will focus on changing patterns of work and family life that affect men's and women's experiences of time pressure.

A major theme of this book is that the rhythm of our lives, the very meaning of work and leisure, is being reconfigured by digitalization. But at this juncture, it is helpful to consider other, often overlooked, dimensions to and causes of harriedness. I want to bring to bear on this question some interesting and credible data about how people actually use their time. Such close scrutiny will reveal the limitation of treating all time as the same, as if we only inhabit one time-space, that of acceleration. It may also help to resolve the riddle of how it is that we often feel we have less time for the things we want to do than we actually have.

How people spend their time matters for quality of life, irrespective of the income generated, as economists and even governments have begun to grasp. An indication of this was when former French president Nicolas Sarkozy set up the Commission on the Measurement of Economic Performance and Social Progress to explore the limits of GDP as an indicator of economic performance and social progress. It concluded that "the time is ripe for our measurement system to *shift emphasis from measuring economic production to measuring people's well-being*."[1] Chaired by Nobel Prize–winning economist Joseph Stiglitz, the commission highlights the fallacy of assessing well-being in terms of financial resources alone. This is part of a wider recognition in economics that people do not necessarily become happier when they become richer.[2]

In a similar but more philosophical vein, *Discretionary Time: A New Measure of Freedom* sets out to change our thinking about what makes for quality of life. According to the authors, how much time we have matters just as much as how much money we have. The book's argument is built around one powerful idea, that being able to choose how you spend your time is central to an individual's sense of freedom:

> When we say that someone "has more time" than someone else, we do not mean that she has literally a twenty-fifth hour in her day. Rather, we mean to say that she has fewer constraints and more choices in how she can choose to spend her time. She has more "autonomous control" over her time. "Temporal autonomy" is a matter of having "discretionary" control over your time.[3]

Conversely, the less you are able to determine how your time is spent, the more your being is "unfree," or deprived. The concept of discretionary time as a measure of freedom is very appealing. It echoes earlier ideas about the desire for temporal sovereignty, or control over one's time, as a significant measure of life satisfaction and well-being.[4] Such notions support normative arguments that treat the allocation and availability of time as important dimensions of social justice and of legitimate political concern. Instead of focusing on speed, per se, to further our exploration of the acceleration society, we should be looking at changing dynamics in the *distribution* of time.

Free Time and Time Pressure

It is no wonder, then, that huge interest has been generated by the idea that time is in short supply in modern societies. Complaints about the twenty-four-hour day being too short to fit everything in are common in the academic and popular presses. The inexorable increase in the pace of life is viewed as a perverse symptom of late modernity, leading to increasing pressure and stress. Understanding why time pressures have increased is a critical social question not least because of the consequences for physical and mental health outcomes.

Sparking this debate was Juliet Schor's book *The Overworked American*.[5] It claimed that from the 1970s through the 1980s, Americans were working longer, and that this applied generally across the spectrum on income and family type. Her argument was echoed by many others, including Arlie Hochschild, whose book title *The Time Bind* entered general currency. The finding that American workers are logging more time at the workplace than their parents and grandparents touched a chord in the popular imagination. Whereas economic progress and increased prosperity were supposed to deliver more leisure time, instead time scarcity and the paucity of leisure time seem to be the result. Media exposure of notions like the "time squeeze" and "time famine" rapidly became part of the folk narrative about the pressure of time in modern life.

A linked concern was whether, as a consequence, parents are spending fewer hours with their children. Most Americans agree or strongly agree to survey questions asking whether "parents today don't spend enough time with their children."[6] The cultural image of the modern mother changed from the devoted homemaker to the frenzied, sleepless working mom. The conventional wisdom accompanying this change is that today's mothers, who juggle the dual roles of worker and family caregiver, spend less time with their children and receive relatively little help from fathers. Social commentators worry about the quality of family life. The politics of time has thus become a major issue that has largely taken the form of a discussion about work-family balance and the quality of contemporary life.[7]

How Much Time?

But how can we measure the pace of life? As a subjective state, an acceleration of the speed of life affects people's experience of time. It causes individuals to consider time to be scarce, to feel rushed and pressed for time. In other words, people feel that they can no longer find time to complete the tasks and activities most important to them.

That time pressure is a common experience is evidenced by the fact that an increasing proportion of the population report feeling short of time. Since 1965, the US time-use researcher John Robinson has been asking adults, "Would you say you always feel rushed, even to do things you have to do, only sometimes feel rushed, or almost never feel rushed?" The proportion of Americans reporting that they always feel rushed rose from 25 percent in 1965 to 35 percent forty years later.[8] Almost half now also say that they almost never have time on their hands. According to most evidence, people perceive leisure time as scarcer and more hectic. And this is also true cross-nationally, where there has been consistent historical growth of busy feelings through the last part of the twentieth century.

Do these widespread perceptions of time pressure reflect the behavioral evidence of how people spend their time? Has leisure time actually decreased?[9]

Let us begin by looking at trends in the hours of paid work. Several commentators have shown that there is surprisingly little empirical evidence supporting Schor's claim that the average length of the workweek has changed appreciably in recent decades. The issue is still a matter of controversy, at least at the margins, as different methodologies yield somewhat different conclusions, and there is significant variation between different countries. In the United States, average hours have held broadly constant for many years, as they have for example in Australia, Finland, and Sweden. In European countries, such as France and Germany, they declined as a result of deliberate government policies designed to reduce working hours.[10] As a result, in France and Germany, employees work only about 80 percent as many hours a year as employees in the United States.

Overall, however, in both the United States and Europe, there has been no straightforward increase in working hours over the last fifty years.

Indeed, between 1965 and 2010, when over one-third of Americans felt rushed, their free time had actually increased.[11] Using data from time diaries, the most direct and reliable methods of measuring free time, this finding has been replicated in multiple surveys across nineteen countries. The long-term growth in leisure for the working-age population is evident in nearly every country for which we have appropriate evidence.

So how do we account for this mismatch with people's experience of a rising deficit of time?

A range of explanations has been proffered to account for this paradox. They all contain partial truths, and indeed are not mutually exclusive, so it is worth considering the contrasting approaches here in some depth. Let us begin by considering those that point to economic change as the root cause of a time squeeze.

One key to this paradox lies in distinguishing between the amounts of time available to different groups of people within a country. While the average workweek has barely changed over the last few decades, the overall trend is one of an increasing polarization of working time, between those who work very long hours and others who work few or no hours.[12] Long hours in some groups are countered by growing numbers of employees working relatively short weeks. The increased dispersion means that the proportion of those with substantial increases in workloads has grown. Critically, long workweeks impinge disproportionately on dual-career families, as both members of a couple with a very long (combined) workweek are more likely to be highly educated and in high-status jobs. The fifty-plus-hours week is thus predominantly a characteristic of the professional and managerial class, those "likely to shape the terms of public discussion and debate."[13]

When theorists of the acceleration society refer to the hastening pace of life, they have in mind the abstract subject. They are not attuned to the detailed manner and circumstances in which time is organized into daily routines by gendered individuals and negotiated within households. Consequently, they fail to see that what happens to an individual's average hours of work is not the same as what happens to work collectively within households. The vicissitudes of scheduling and the intricacy with which our lives are tied to others can only be fully understood by treating the household, rather than the individual, as the unit of analysis.

One of the greatest social changes of the second part of the twentieth century has been the widespread participation of women in the workforce. While the contribution of men has significantly declined in the UK, the United States, and most industrial countries, the hours that women (especially mothers) contribute to the labor market have significantly increased.[14] This has led to dual-earner families outnumbering male-breadwinner families. Today, roughly 60 percent of two-parent households with children under age eighteen have two working parents.[15] The generalization of dual-earner couple is not limited to the United States but is a trait of every economically advanced country. Discussions about average working hours thus mask a dramatic redistribution of paid work between the sexes.

It is as if much of the paid work has been transferred from men to women. The resulting dual-earner households are supplying more working hours to the labor market than ever before. Time pressure is especially strong in families with dependants, where both husband and wife are in full-time employment. The widespread perception that life has become more rushed, therefore, has as much to do with real increases in the combined work commitments of family members as it is about changes in the working time of individuals. "What, for more than a decade, has been taken to be a controversy about overwork (i.e., trends in individual hours) . . . is actually a manifestation of the difficulties of reconciling (paid) work and family responsibilities, following the historical demise of the male breadwinner model."[16] This transformation in family composition and gender relations is central to explaining the time deficit.

In order to understand our experience of living in an *acceleration society* then, we need to consider how households are organizing their working and nonworking lives and be attuned to gender differences in time pressure. Feminist scholars have long argued that the squeeze placed on women's time is due to combining paid employment with their responsibility for the household's operation.[17] Indeed, time-use data suggest that time poverty is a particularly widespread experience among working mothers, who juggle work, family, and leisure.

Unsurprisingly, single mothers feel the worst about the time they allocate to their children and their numbers have expanded greatly since the 1960s.[18] The breakdown of marriages involving children can radically exacerbate time pressure if a lone parent is forced to serve as both breadwinner and homemaker. When children are reared outside a two-parent home, fathers are much less likely than mothers to shoulder the day-to-day responsibility of caregiving—that is, fathers are far likelier than mothers to drop the parental role altogether. Mothers cope by reducing their hours of paid work, especially when the children are young, in order to perform the household work and child care. As lone mothers are overrepresented among the poor, they are also unlikely to be able to buy paid domestic help. Lone parents, then, have much less discretionary time than dual-earner couples, with or without children.

Time Spent with Children

For people with children, spending time with them is regarded as one of the most desirable uses of discretionary time. We want both enough time and "quality time" with our children. Let us first consider the amount, as there is a widespread view that time pressure is squeezing out precious time with children. For example, Hochschild claims that family time is being crowded out by long hours of paid work. In fact, time-use data show that both mothers and fathers are spending more time with their children than ever before.[19] Although there are variations between countries, overall, parents are averaging more time with their children, despite working longer hours. So how is this possible?

This aspect of the time-pressure paradox is addressed by Suzanne Bianchi, John Robinson, and Melissa Milkie in *Changing Rhythms of American Family Life*: "although parent-child time has remained steady or increased over the years, almost half of American parents continue to feel they spend too little time with their children."[20] The explanation, as we shall see, centers on the cultural ideals of intensive parenting combined with the nostalgia for a mythical past of more quality family time. But let me first report their extensive findings on feelings about time.

Among employed mothers, almost half (47 percent) feel they spend too little time with their children, whereas only 18 percent of nonemployed mothers report this. Working mothers are also the most likely to feel most time pressured and that they are constantly multitasking. Married fathers are significantly less likely to feel "always rushed."[21] Where the authors found very large differences was between mothers and fathers in feelings of "too little time" for oneself. Some 57 percent of married fathers and 75 percent of employed mothers expressed this. And whereas married mothers craved more time alone and with their husbands, married fathers wished they had more time with their children.

How, then, have parents preserved their time with children? A common explanation is that much housework has been mechanized. We will consider this argument in chapter 5. Suffice it to say here is that, despite dramatic improvements in domestic technology, the amount of time spent on household tasks has not actually shown any corresponding dramatic decline. However, there has been a marked shift in the composition of time away from routine domestic work (cooking and cleaning) and toward child care but also shopping and odd jobs. But the total time in unpaid domestic labor of all kinds held constant over the twentieth century.

The major change has come from working women themselves, who reduce their time in unpaid labor at home as they move into the workforce. However, they do not remotely reduce their housework hour-for-hour for time spent in paid labor. And while their male partners increase their own time in housework, this is not nearly as much as working wives reduce theirs. The upshot is that rather less unpaid household labor gets done overall in the dual-earner household—but women's total combined time in paid and unpaid household labor is substantially greater than is the typical nonemployed woman's in domestic labor alone. Working mothers' combined time in paid and unpaid household labor typically exceeds fathers' by five hours a week and nonemployed mothers' by nineteen hours a week.[22] The working woman is much busier than either her male colleagues or her housewife counterpart.

Moreover, it is also primarily women who adjust their working hours in relation to the number and ages of their children. The decrease in women's employment comes with the birth of the first child, and they continue to curtail their hours after they have children. Fathers, if anything, tend to increase hours of employment after the birth of a child. Men and women thus make different decisions in allocating their time, based in part on choice and in part on institutional forces and cultural pressures. These differences have diminished in recent years, but mothers continue to adjust their schedules more than fathers do. As a result, there is still a pay penalty associated with motherhood, whereby mothers' wages lag behind fathers' in virtually all developed countries.[23]

That women fit themselves to their family shows up in their time strain; married women want more time to themselves because they have adjusted to being there for their children. Cultural expectations about what mothers and fathers should do remains strong, with mothers feeling more of a need to put children and family first as they sacrifice their own need. Overall, it appears that parents are giving themselves over to rearing children to the extent possible given other demands on their time and limited resources in some families. Yet they feel as if their efforts are not enough.

As noted above, these feelings of time pressure are largely the result of normative changes in expectations about good parenting.[24] Among both working-class and middle-class mothers, good mothering is defined in terms of devoting unlimited time and resources to their children. As family size becomes smaller, children become a central focus and are seen to require extraordinarily labor-intensive parenting. Intensive mothering is a cultural ideal to which women are expected to sacrifice careers, leisure time, and whatever else is necessary to ensure that their children thrive.

Fathers are expected to be equally involved in care giving. The last half-century has witnessed dramatic changes in men's attitudes toward parenting and in conceptions of masculinity more broadly. However, the shared belief in egalitarian family relationships has not yet been matched by men's behavior. For example, the substantial increase in the time fathers spend with children is three times larger on weekends than workdays, with the result that routine child care responsibilities are left to their spouses on days when they must work.[25] And, rather than fathers replacing mothers' time, mothers are present for most of the time that fathers are caring for their children (see chapter 5). While this family time spent together may well be the most cherished, it does result in a gender gap in leisure, as I argue below.

Finally, women's role as household manager also adds to their feeling of always feeling rushed. Being responsible for managing something as complex as children's lives and a home, even when away from home, "may also account for the large gender discrepancy between fathers' and mothers' feelings about needing more time for oneself, feeling rushed, and feeling like they do more than one thing at a time."[26] Mothers' greater subjective sense of time pressures may derive from their being the one who continues to orchestrate family life—a reality that is difficult to capture in linear time-diary data.

Such arguments involve moving beyond quantifying the volume of (paid and unpaid) hours worked in order to broach the more subtle, qualitative aspects of the meaning and experience of time. They suggest the need to explore the ways in which time is ordered and practiced as well as the density or intensity of the lived experience of time. It is clear that time pressure is complex and multidimensional. Below I will outline a schema that delineates three different mechanisms that cause time scarcity. But first let us turn more broadly to the cultural connotations of being busy.

Cultural Acceleration: Busy Lives

So far we have examined a range of economic and demographic factors that contribute to time pressure, such as changes in the labor market, working hours, and the composition of households. We have also considered how contemporary discourses of hyperparenting heighten perceptions of time scarcity. But there are another set of explanations that primarily focus on consumption. These explanations are related in that the dramatic shift in women's employment coincided with the "overwork culture," turning workers into "willing slaves" who are prepared to work ever-longer hours in a society that equates busyness with success and status.[27]

Such arguments see consumption in purely negative terms, as fueling long working hours because of our competitive consumerist culture. According to Schor, we are trapped in a "squirrel cage," an insidious cycle of work and spend where we compete with our neighbors' lifestyles and compensate for our lack of time with children by buying them things. Why, she asks, don't affluent North Americans "downshift" and reduce both their hours of work and levels of consumption as a way out of the work-spend cycle?

If only things were so straightforward. There are deep-seated psychological reasons for apparently unnecessary consumption in a capitalist society in which people's sense of self and sense of freedom come to be defined by money and possessions. For a long time social scientists have talked about the complex relationship between shopping and individual identity formation and the extent to which purchasing goods is a social practice oriented to others.[28] The requirement for individuals to narrate their own identity through styles of consumption brings with it the demands of trying new and varied experiences, and this leads individuals toward the insatiable pursuit of more cultural practices. In short, being busy has become a necessary condition of a fulfilling lifestyle.

Perhaps cultural discourses that value action-packed lives, coupled with high levels of consumption, are to blame for upward-spiraling perceptions of feeling rushed. Indeed, busyness may result not only in stress but, for some, in feelings of increased happiness or life satisfaction arising from the positive energy connected to states of arousal.[29] Such an approach reformulates the discussion about the socioeconomic correlates of time pressure into a debate around the manifestations and consequences of busyness. "Whereas the concept of 'time pressure' is negative in its connotations, 'busyness' is at worst neutral, and may indeed carry with it the positive connotations of 'busyness' as an antonym to 'idleness.'"[30]

So has the notion of busyness acquired a new positive meaning in our culture? Is busyness a status symbol for those with higher social capital? In an intriguing argument, Jonathan Gershuny claims that whereas a century ago those in the upper income bracket were defined by their leisure, in a reversal of Thorstein Veblen's classic *Theory of the Leisure Class*, nowadays prestige accords to those who work long hours and are busiest at work.[31]

To this point I have described how the increase in dual-job households in conjunction with shifting norms of parenting contribute to time pressure. Entirely consistent with this explanation, however, are two further arguments. One is to do with the density of leisure itself, created by the desire for ever more intensive consumption of goods and services.[32] The second involves not so much a change in behavior but a change in the way feelings of "busyness" are constructed out of these: "the growth in busy feelings may in part reflect an increasingly positive view of 'busyness' that results from its association with the increasingly busy lifestyle of the most privileged groups in developed societies."[33] Today, it is conspicuous devotion to time-intensive work activities rather than the conspicuous consumption of leisure that is the signifier of high social status.

"Busyness" is a subjective state that results from the individual's assessment of his or her recent or expected activity patterns in the light of current norms and expectations. However, for busyness to be an externally observable behavior, it would need to be reflected in long hours of paid work and in the density of work and leisure, that is, the frequency and variety of activities undertaken. (It would also be evident in the multiplicity of simultaneous activities, a topic not covered here but which, I will argue in subsequent chapters, is key in the digital age.)

Gershuny finds little evidence for such objective behavioral changes in busyness. While higher-skilled groups did increase their paid work time relative to lower-skilled groups, paid work overall declined for both men and women. Significantly, there was no increase in the *intensity* of activities (on either a workday or a nonwork day). Although this empirical evidence cannot prove that there has been a change in the social construction of busyness, Gershuny concludes that it is consistent with his argument. If there has been no behavioral change, then the explanation must lie in the changed cultural meaning of busyness. One part of the resolution of the time-pressure paradox, then, is that busyness, and not leisure, is the "badge of honor."

According to this view, then, busyness is largely a cultural orientation. Certainly, the argument resonates with the representation of some groups, such as financial traders and corporate executives, who are invested in high-pressure, burnout careers and whose status derives from their workaholism. However, the parallel with Veblen's leisure class, while striking, exaggerates the freedom of this new superordinate class whose work pressures are largely the result of managerial performance measures. They may embrace the high-speed work culture, but it is important to emphasize that it is not entirely of their choosing.

This is particularly pertinent given the extent to which, over the last decade or so, firms have sought to convert many high-wage and full-benefit workers to contingent and contract workers. The managers and professionals who have survived work longer hours to ensure their job security and increase their chances of promotion. Long hours are the principal means of demonstrating commitment and ambition to employers. A concurrent trend in the economy has been the expansion of work schedules into evenings and weekends. The profound influence of mobile technologies on working time is the subject of subsequent chapters. At the very least, discussions about the symbolic status of busyness should take into account changed employment conditions and concerns about job security.

Moreover, the culture of ostentatious work performance is one in which men can more easily immerse themselves than can women. While there has been a convergence in women's and men's aspirations for high-powered managerial or professional careers, my own research on corporate managers reveals that the domestic circumstances of women differ from those of men.[34] While a significant number of male senior managers have partners who are not in paid employment, women managers generally live in dual-career households. Women are therefore more likely to experience intense friction between the demands of career and family life. By overlooking this issue, Gershuny leaves us with the impression that the new world of work is gender neutral. The changing norms of busyness are a vital element in resolving the time-pressure paradox, but we must be wary of the implication that it is equally seductive to all.

If busyness in paid work can be a form of status distinction, so too can busyness in leisure. Reexamining Gershuny's argument, Oriel Sullivan points to the often overlooked importance of the density of leisure. It turns out that those who work long hours in employment also have a greater leisure density.[35] In other words, they ceaselessly pack more and more activities into the same time period. Otherwise, how can income-rich, time-poor households in affluent Western economies both work more and consume more? Her answer identifies two *temporal strategies of consumption*: leisure consumption can happen faster and goods may be continuously replaced with more expensive alternatives. These are both strategies for maximizing the "time yield" in time-pressured modern societies.

There is a vast literature on the nature of consumption and the cultural tastes of modern consumers. However, little is known about the pace or busyness of leisure participation. To measure the "voraciousness" of leisure consumption, Sullivan analyzed the frequency of five out-of-home leisure activities: going to the cinema/concert/theater; eating/drinking out in a restaurant, café, or pub; playing sports/keeping fit/walking; watching live sports; and attending leisure activity groups.[36] The logic of this list is that these activities take both time and money to engage in and require a degree of temporal planning and coordination. Indeed, she found that high-status, dual-earner couples with dependent children use both temporal strategies of consumption the most. They have the highest level of participation in these leisure activities and they continuously upgrade their consumer goods (without the time to use them).

Once more, gender and social status reinforce each other so that the greatest differential in voraciousness is between men with high social status and women with the lowest. Embracing a busy, diverse pattern of cultural consumption practices has thus become a mark of distinction among high status groups.

To date, most literature on time pressure has focused on the impact of working practices, both in the sphere of employment and in the domestic sphere. In modern, time-pressured societies, the pace of leisure is also of signal importance. I will return to this topic below, and later I will be looking specifically at how ICTs intensify leisure. But first, such arguments about the busyness of work and leisure time point to the difficulty and complexity of measuring the temporal rhythms of daily life. Not all activities have the same tempo. Time as measured by the ticking of the clock cannot remotely capture our quotidian experience of multiple and overlapping temporalities. If we feel short of time, it can be for a variety of reasons and take a variety of forms. Indeed, it may even be the case that some of us have more time, but not time of the right kind or when we need it.

The Temporal Disorganization of Daily Life

It is for this reason that I turn to the work of Dale Southerton and Mark Tomlinson, who argue that the experience of harriedness is multifaceted, depending on which aspect of temporality is being "squeezed."[37] They distinguish three mechanisms that generate different senses of feeling pressed for time. First, the *volume* or duration of time required to complete sets of work and consumption tasks is the basis for the *substantive* sense of being harried; second, *temporal disorganization* is the outcome of the difficulties of coordinating social practices with others; and third, *temporal density* accounts for experiences of time that can be described as *juggling* and *multitasking*, that is, the allocation of certain practices within temporal rhythms that create a sense of intensity in the conduct of those practices. This attempt to delineate different aspects of time pressure captures well the idea that harriedness is a multidimensional experience.

Up to this point, we have focused on the first dimension, pressures derived from the volume of (clock) time available or the substantive sense of being overloaded. The most comprehensive account of this, as we have seen, is provided by time-diary studies that record and measure the number of minutes devoted to different activities. Diaries can be employed to account for patterns of convergence and divergence of time use across different countries, social classes, and gender, and to explore the domestic division of labor.[38] However, the duration of activities remains the primary focus of analysis. They are much less fit for the purpose of analyzing qualitative dimensions of time, such as the tempo or intensity of activities, and the rhythms and sequences in which activities are conducted.

Here I want to consider the second mechanism listed above, *temporal disorganization*, which highlights the importance of looking at the rhythms or organization of life. This sense of harriedness is less conspicuous than the substantive form, because it accounts for experiences that are not obviously connected with an absolute shortage of time. It is particularly apposite, however, given the extent to which collective social practices, derived from institutionally stable temporal rhythms, have been eroded.

We live in a society in which the standard working week, where work was synchronized for a substantial proportion of the population, is no longer the norm. Flexible working hours, 24/7 working time, and contract work create coordination problems, as working times and locations are increasingly deregulated and scattered. The growth of nonstandard evening and weekend work hours is also associated with increased time pressures, in part because they decrease individuals' abilities to mesh work schedules with the social activities of friends and family as well as to find adequate time to sleep. While higher socioeconomic groups may be able to utilize flexibilization to gain greater control over their time, lower status groups suffer from temporal fragmentation caused by working irregular hours.

Nevertheless, socializing is made more difficult for both groups because of the weakening of the shared sociotemporal order and a corresponding fragmenting of activity. As Southerton elaborates, whereas the middle classes tend to meet more by prearrangement, the working classes use public spaces to meet where there is a strong likelihood of meeting network members by chance.[39] In both cases, however, coordination becomes increasingly problematic. It means that by "turning up" in public spaces, one is less likely to meet friends and acquaintances because they might work at different times of the day. (No wonder mobile phones have become essential aids to the personalization of scheduling—a matter we leave for discussion later.) Various consumption practices also involve interaction within social networks that require coordination. In sum, the problem of coordination is collective—it requires the alignment of practices across the schedules of social networks. In this sense, harriedness is a consequence of the difficulty of coordinating practices in time and space.

A good example of this is provided by the emergence of convenience food. The widespread use of partially and totally prepared food reflects the desire to save time, as those with most money and least time buy more of it. Such consumer products are widely seen as solutions to the problem of cooking and eating in the context of a busy, time-pressured lifestyle. However, Alan Warde argues that they are as much a response to the reordering of the time-space relations of everyday life as a modern search for the reduction of toil.[40] While family meals retain their symbolic significance as important forms of sociality, they now pose considerable scheduling problems in getting people together in the same place at the same time. The cook has to be in the right place for long enough, but so do those for whom she cooks. The erosion of institutionally fixed routines and the fragmentation of daily activities mean that more negotiations, more decisions, and more effort are required to perform the necessities of daily life. For a significant proportion of people, planning to meet people becomes a major preoccupation.

According to these arguments, underpinning much of the feeling of haste is the difficulty of synchronizing time-space paths: "the problem of timing supersedes the problem of shortage of time."[41] People may have more time, but more important than the hours worked is their dispersal through the week. While acknowledging that the increase in women working does contribute to time pressure, the emphasis here is on the intransigent problems of scheduling in a deroutinized society. Discussions of harriedness have not paid sufficient attention to the aspect of space and the need for interpersonal conjunction.

While this exposition of temporal disorganization adds a great deal to our understanding of time pressure, the focus has again shifted somewhat from households toward individuals. We have already seen that dual-earner families are the most time pressured, and it is precisely at the level of households where most of the coordination problems arise. The issue of coordination among spouses is a crucial underpinning of the demand for flexible start and finish times at work. When both partners participate in the labor force, the family's day becomes more complex as work schedules may not overlap or, in other words, may be desynchronized. Everyday family life is different when work schedules are desynchronized, as couples tend to spend less time together but share domestic and parental work more equally.

The incidence of, and reasons for, off scheduling among dual-earner couples (that is, partners' timing their work schedule so that one is not working when the other one is) are therefore directly related to time pressure. And, indeed, there is strong evidence that off scheduling is both widespread and growing. Based on a study of French time-use surveys, Laurent Lesnard found that families in which both partners work standard workdays (of approximately eight hours) represent less than half of total family workdays.[42] About 70 percent of the work time of these couples is simultaneous (synchronous). For the rest, atypical family workdays are the result of long hours, shift work, part-time work, and short or irregular work hours. This pattern is consistent with findings for the United States and many other modern economies.

But is this trend the result of couples freely choosing to reschedule their work, trading time together for parental efficiency? In the study, the few couples who have control over their timing overwhelmingly favor family workdays, suggesting a strong general preference among couples for synchronized work schedules. However, most couples do not have temporal autonomy over their work hours. Atypical work schedules and off scheduling are highly correlated with employment sector, occupation, and position on the social ladder. When husbands' positions are at the management level, standard workdays are much more common than for factory-worker families. Those who have no power over their work schedules have strikingly higher chances of having desynchronized family workdays. In other words, very few couples are free to fix the timing of their work hours. Dual-earner couples' off scheduling is thus a consequence of employers' economic behavior rather than employees' preferences.

The rapid rise of family time over the last decades indicates that being together is of increasing importance for the contemporary family. However, less temporal complementarity between parents is negatively correlated with partners' time together and with children. The prevalence of off scheduling therefore adversely impacts on family solidarity. It points to the difficulties parents face as they seek to balance market work and family. As Lesnard notes, the UK's flexible working time law (in place since 2003) was one of the first that obliges employers to consider employees' requests for more family-friendly work schedules. Such policies, and those that limit long working hours, might begin to address the problem of time pressure. I will be exploring the wider politics of working time in my last chapter.

Temporal Density

Let us finally turn to the third mechanism that generates feelings of time scarcity, *temporal density*. This notion accounts for experiences of time that can be described as *juggling* or *multitasking*, that is, the allocation of certain practices within a given period of time. Allocation refers to certain practices being located within temporal rhythms that create a sense of intensity in the conduct of those practices. It is directly related to temporal disorganization, but here the emphasis is not so much on sequencing, but rather highlights the simultaneity of different activities or multitasking. Allocation is also linked to a notion of boundaries that separate practices. The allocation of practices, which no longer have clearly defined boundaries, into particular parts of the day can generate senses of being harried, irrespective of whether the bulk of that day is experienced as being "pressed for time."

Southerton and Tomlinson illustrate this point by drawing on qualitative interviews with twenty British suburban households. For example, although Chloe said she did not feel "generally pressed for time" in the survey, she said she felt busy "some of the time but not others":

> Some mornings are chaos, after getting them off to school I'll have a cup of tea and a sit down, then I'll try and get all the housework done so that I can get off to work for 12.00 and that's as busy as getting the kids off, you know, start the washing, do some ironing, make the beds, then the washing finishes, so I stop what I'm doing and peg it out . . . . Work is easy, the most relaxing part of the day because I only have to do one thing.[43]

Here we see the harried experience of temporal disorganization, as Chloe rushes to meet the institutionally fixed school meeting times, as well as that caused by the temporal density of housework. Such examples proliferate in the feminist literature about how women in dual-earner families constantly juggle their work and households tasks.[44] The fear has been that working women will simply add a "second shift" of paid employment to their existing responsibilities for housework and child care. This lies behind the suggestion that women are suffering from time poverty. There is now talk of a gender gap in leisure.

We began this chapter by making a case for regarding discretionary time as an important measure of freedom and equality. Indeed, Nancy Fraser has argued that gender equity needs to be reconceptualized as a "complex notion comprising a plurality of distinct normative principles."[45] One of the seven principles that she proposes as crucial to gender equity concerns the distribution of leisure time. This issue is more complex than it first appears. Taking into account temporal density leads to a reformulation of the gender gap in leisure. The crucial issue is not only that women may have less leisure time, but that women's leisure time may be qualitatively "less leisurely" than men's.

So do women, as is commonly claimed, have comparatively less leisure time than men? In the United States, employed mothers' combined time in paid and unpaid work exceeds that of fathers, as noted earlier. Surprisingly, however, this difference is not substantial. Analyzing time-use data from ten countries that are members of the Organisation for Economic Co-Operation and Development, we found that, when taking paid and unpaid work together, there is little difference in the number of minutes men and women spend in "work."[46] Unpaid work (housework, child care, and shopping) continues to be highly gender specialized in that women's share is approximately three-quarters. But despite this, the quantity (the number of hours) of leisure time that men and women have is remarkably similar.

How can we reconcile the apparent gender equity in the objective quantity of leisure with the subjective impression of increased time pressure among women? In my view, the answer lies in considering the density or quality of leisure time available, and not simply the quantity. For example, the tendency to perform multiple and overlapping tasks simultaneously creates a distinctive experience of leisure. It would be seen as a form of work intensification that increases productivity if it were paid. Such evidence as we have suggests that multitasking has greatly increased over the last quarter of the twentieth century.[47] This could be taken as an objective measure of the acceleration of the pace of life.

We must therefore distinguish between degrees of leisure. In popular discourse, leisure is conceived of as free time, time at one's disposal, or *pure leisure*. Such leisure with no distracting accompanying activities to constrain it is different from a leisure activity that is accompanied by a constraining activity. Using detailed Australian time-diary evidence, I was able to measure and compare periods of pure leisure, when the primary activity reported is a free time activity with no secondary activity, with periods of *interrupted leisure* contaminated by a simultaneous secondary activity.[48]

This aspect of the lived experience of leisure is at the heart of feminist commentary on the gendered nature of leisure. Analyses of multitasking have usually focused on combinations of domestic labor tasks, such as ironing while watching television. I also analyzed the combination of parental leisure time spent together with children (that is, free time with children). The significant growth in this kind of leisure is one of the main ways that parents have managed to spend more time with their children.[49] But what are the consequences of this for the character of leisure?

One could argue that the best leisure is achieved when playing with one's own children. However it is my contention that this may involve a trade-off when it comes to the quality of that leisure time. The enjoyment of leisure that is combined with other activities might be lower for the parent or the child or for both. The fact that parents derive considerable pleasure from attending to their children's needs does not detract from the argument that they may, at the same time, be experiencing an adult leisure deficit.

The results of my study reveal that men have many more hours of pure leisure, undiluted by unpaid work. More than 60 percent of men's leisure is enjoyed with no accompanying activity. By comparison, little more than half of women's leisure is pure leisure. In addition, men's leisure is less likely to be interrupted than women's. Women's leisure, by contrast, tends to be conducted more in the presence of children and subject to punctuation by activities of unpaid work.[50] The average maximum duration of episodes (blocks of time) of pure leisure is also longer for men. Women's leisure is significantly more harried than men's in that it is more fragmented into periods of shorter duration. In sum, men do have more high-quality leisure time than women.

Overall, this research suggests that there continues to be a gender gap in leisure.[51] The fragmentary character of women's leisure changes its quality. Interrupted leisure, snatched between work and self-care activities, is less restorative than unbroken leisure. It is likely that this leisure will be experienced as more harried and therefore increase self-reported stress. Indeed, it may well be that the contemporary view of increased "time pressure" has more to do with this fragmentation that with any measurable reduction in primary leisure time. The key role of mobile technologies in causing interruptions, with the incessant pinging of phone, text and e-mail messages, will be taken up in subsequent chapters.

This argument is particularly important as a corrective to the focus on the intensification of paid work that has preoccupied industrial relations and sociology of work scholars. Housework is unpaid and never done, resulting in highly gendered time frames that do not coincide with the standardized time of paid labor. Caring labor does not straightforwardly operate according to clock time and cannot be accelerated. "Clearly, the direct activities parents engage in with children consume far less time than the responsibility for overseeing them."[52] While some aspects of care can be commodified and outsourced, the character of intimate personal relationships and emotional labor demand quality time. Indeed, time-use data suggest that working parents who make use of nonparental child care do not reduce their parental child care time on an hour-for-hour basis.[53] Instead, parents, mainly mothers, compress their domestic labor time, squeeze their personal care time, and reschedule the times when they are together with their children so as to preserve their time with children. Perhaps we should be giving as much attention to the intensification of parenting as to the intensification of work.

Conclusion

This chapter has considered a range of arguments about the nature and causes of time pressure, an issue made all the more important by the centrality of time sovereignty to equality and social justice. We have seen that time is multidimensional and is experienced differently by diverse groups of people in modern societies. Extensive analysis of time-use research does not reveal straightforward patterns or causal connections. Rather, it shows that to fully account for the time-pressure paradox, a complex interplay of demographic and economic factors needs to be considered. Key to this are the major repercussions that women's entry into the labor market and the rise of dual-earner families have had for how households are organizing their time.

Acceleration as a phenomenon, then, cannot be understood as a uniform experience of the shortage of time. The many people who are highly time stressed have combinations of characteristics—they have full-time hours of market work, they have family responsibilities, and they are women. Time pressure tends to be especially acute for women whose time is disproportionately constrained by domestic responsibilities. And this is in a context where more and more value is placed on "quality time" with children. Mothers cope by combining leisure time with looking after children, making women's leisure time less leisurely than men's.

Such multitasking, or intensive use of time, exposes the limits of conceptualizing time pressure solely in terms of the volume or duration of (clock) time available. That is why I have emphasized the character of time by looking at issues of temporal disorganization and temporal density. Time scarcity may result not so much from a shortage of time but because of the increasing complexity of scheduling personal, domestic, and work activities. Taking this approach reveals that the experience of harriedness is not uniform, but takes a variety of forms. Time pressure may be experienced by employed lone mothers, by two-job parent families who work shifts, and by dual-career managerial or professional couples, but the sense of time squeeze and the mechanisms causing it differ. The extent to which money can be deployed to alleviate time pressure also differs between these groups.

This is not to discount the cultural significance of our current orientation to, and valorization of, the fast-paced, full life as the good life. The tone of the debate on time pressure has been largely negative, concerned with problems such as balancing work and family life. However, the allure of metropolitan speed is indissolubly linked to the dominant ideals of modernity. Cultural acceleration, doing the most with the time one has and realizing as many options as possible from the vast possibilities that the world has to offer, is the secular version of human happiness.[54] Cultural discourses that value action-packed lives with high levels of consumption lead people to pursue busy lives that are at once both stressful and affirming. The increasing salience of consumption for modern identity formation fuels the demand for faster, hyper consumption. For those within high-status groups, prestige attaches to those who work long hours and are busiest at work. Moreover, these same busy people also have the most voracious pattern of leisure consumption.

The studies I have considered take the important step of bringing cultural norms about time to bear on the paradox of harriedness. However, they run the risk of implying that people are utterly seduced by the celebratory culture of speed in every aspect of their lives. In fact, there is as much organizational constraint as individual choice involved in lengthening working hours, for example, itself a cause of stress. Moreover, a wholesale condemnation of consumer culture, as expressed by Zygmunt Bauman's sentiment that "the 'consumerist syndrome' is all about speed, excess and waste" both exaggerates the place of consumption in modern culture and fails to acknowledge the satisfaction that can be derived from it.[55] Understanding the prevalence of time pressure requires a careful examination of the objective circumstances that cause harriedness, as well as an appreciation of changing cultural norms of busyness.

However, our interaction with the proliferating technologies of acceleration is also fundamental to the contemporary experience of speed. A notable omission from much of the literature on the time squeeze is the role played by information and communication technologies. Modern existence presupposes a technologically replete, multimodal world of ubiquitous connectivity. Electronic communication and new media systems increasingly constitute the fabric of our everyday lives. While there is research on multitasking with children, the study of multitasking with digital technology is only emerging. Yet the concept of the acceleration society outlined earlier presumes a direct link between technological acceleration and the growing scarcity of time. It is to this theme that I turn for the remainder of the book. The next chapter explores these issues in the context of paid work.

Chapter Four

Working with Constant Connectivity

To be doing something, to move, to change—this is what enjoys prestige, as against stability, which is often regarded as synonymous with inaction.

LUC BOLTANSKI AND EVE CHIAPELLO, *The New Spirit of Capitalism*

We know surprisingly little about whether and how information and communication technologies influence the experience of work in the twenty-first century, the subject of this chapter. Yet, the idea that ICTs are intrinsically technologies of acceleration is particularly compelling in relation to paid employment. Both the popular press and the academic literature trumpet their potential to save time, speed up work, and allow people to work anywhere at any time. Techno-enthusiasts encourage us to think about work and organizations in new ways and imagine "virtual organizations," "flat worlds," and the ability for work patterns to "follow the sun" and "span the globe." Exactly how this will occur is less often discussed, but all agree that these technologies overcome the traditional temporal and spatial constraints of work. Perversely, in this new world of work, the more rushed we feel, the more we turn to digital devices to relieve the time pressure.

So how can we begin to unravel the relationship between the two essential dimensions of work: time and technology? Work is done in time; it is a temporal act. At its most basic, the employment contract represents the exchange of pay for the capacity to work over a specified period of time. However, as the activities of work become more complex and heterogeneous, it is harder, even for managers, to measure work in terms of the length or quantity of time: minutes, hours, days, weeks. We know that clock time cannot capture the properties of time, and that how we experience it has to do with social norms and organizational rules, conventions and culture.

The process whereby labor power (or "human resources") is transformed into a product or service always involves tools and techniques. Examining how technological innovations enhance productivity by increasing the pace of work has been the focus of numerous studies. The iconic case is Henry Ford's assembly line, which became standard in automobile plants all over the world. The advent of digital technology, however, represents a major and distinct break from earlier generations of machine technology. Crucially for my argument in this chapter, employees using information technology in modern work situations are largely responsible for the pace and rhythm of production.

This chapter will consider the temporal implications of ICT for men and women workers and ask whether the explosion in their use is to blame for our feelings of time pressure. Associated questions to explore are whether the stress of modern working life is attributable to the constant digital chatter that envelops us and the tendency for working time to colonize personal, family, and leisure time. More profoundly, are the temporal rhythms of work speeding up or assuming a radically different character?

Positive and negative outcomes are commonly associated with the anytime/anywhere connectedness of multiple devices, such as flexibility, mobility and enhanced work-life balance, along with the increasing pace of work. However, the received wisdom is that the pervasiveness of ICTs is intensifying work and fragmenting the workday, as well as extending work's reach.[1] I will take issue with this, arguing instead that work practices are being reshaped as employees negotiate the constant connectivity intrinsic to their work. In doing so, I will expose as simplistic the notion of the technologically tethered worker with no control over their own time. It fails to convey the complex entanglement of contemporary work practices, working time, and the materiality of technical artifacts.

This leads me, in the concluding section, to consider the relationship between time, technology, and the meaning of work for identity. I will examine whether the wider socioeconomic conditions of work, rather than the technologies themselves, are producing acceleration—in particular, whether changes in the nature of careers and organizational cultures are contributing to feelings of anxiety about speed. The disruption of linear time as it was lived according to the standard narrative of long-term, stable employment also crucially affects our perceptions of the pace of work.

Looking Back in Order to Look Forward

Where better to start than with some history, including my own? It was as a student of industrial sociology in the 1970s that I first became interested in the relationship between technology and the speed of work. As I outlined in chapter 2, historians of the Industrial Revolution have written about the importance of clock time in disciplining labor and increasing productivity in early manufacturing. The emergence of the automated car assembly line was the culmination of this drive to apply technology to intensify the rate of work. Henry Ford's renowned innovation built on the organizational principles of scientific management. Taylorism increased productivity by breaking down the labor process into component motions and work tasks, according to rigorous standards of time and motion study. Incorporating these principles, Fordism achieved dramatic gains in productivity by flowing the work to a stationary worker. The efficiency of this system of mass production was based on the subordination of workers to the momentum of the machine.

An avid participant in what became known as the labor process debate, I learned that technology is central to the control of work in a capitalist economy.[2] In particular, this literature provoked my interest in the design of technology, and its influence is especially evident in the first edition of *The Social Shaping of Technology*.[3] Rather than conceiving of technology as an autonomous force determining the organization of work, I argued that technology is itself crucially affected by the antagonistic class relations of production. While this perspective was reductionist in its own way, seeing technology as a tool of management, it did direct attention to the social relations of technology.[4]

Forensic studies of the evolution of technologies revealed that patterns of power and cultural values shape the actual processes of technological development. For example, David Noble's famous study of the automation of machine tools showed that automation did not have to proceed in the way it did; rather, the form of automation was the result of deliberate selection. Industrial innovation is, then, a product of a historically specific activity carried out by social groups for particular purposes: "behind the technology that affects social relations lie the very same social relations."[5] In other words, technical choices are simultaneously social through and through.

Moreover, Noble stressed that there were many different levels of social determination of technology. While management's demand for control over workers was critical, so was the role of the military, as well as the ideology and interests of engineers. Importantly, he argued that "engineering rationality" takes the view that the most efficient production process is the most automated. This impulse, to eliminate the human element from production because it is the potential source of "human error," still prevails.

In brief, the social shaping approach was concerned with emphasizing that politics and negotiation are the key processes through which technical possibilities are, or are not, put into practice. That is, the technologies of production we have are in some sense a reflection of the social relations of production. Our technological infrastructure and the material configuration of specific devices create more or less favorable conditions for job autonomy, control over effort, and the parameters of working time.

It is worth recalling, however, that this focus on design does not imply that the makers or marketers of a technology can reliably predict its final use. According to the textbook model, innovation is an activity restricted to engineers and computer scientists and, thereafter, technologies are autonomous and permanently fixed. In reality, studies have repeatedly shown that the way people adopt technologies is not necessarily in line with their intended usage.[6] Nor are the affordances of technologies utilized in the same way by all users. Treating technologies as sociomaterial practices allows us to see that long after machines leave the industrial laboratory, their formation is still taking place. Technological change is thus a thoroughly contingent and heterogeneous process. There is nothing inevitable about the ways technologies evolve and are used in daily temporal practices.

This point is particularly salient in relation to digital devices. The mobile phone and the personal computer are inherently more flexible than older, industrial technologies. While the utility of a machine tool is limited, a smartphone, for example, combines a phone, text messaging, e-mail, a web browser, camera, and a video recorder—and, importantly, it is all these things at once. So the device is not just one thing, it is what people make of it and how it connects to the existing social dynamics of work. In other words, we need to examine how users interact with ICTs in particular organizational contexts. So when I talk about how ICTs are changing the temporal rhythms of work, I always mean ICTs as they are embedded in social institutions, with preexisting conventions, cultures, norms, and objectives.

The Networked Worker

There is no denying that computerization has dramatically transformed work, arguably as radically as the industrial revolution did. While the shift to a postindustrial knowledge economy is often overstated, the effect of information technology on the structure of employment and the nature of jobs has been immense. The rise of service and administrative sectors means that our old paradigm of the technology of production, based on an age in which manufacturing dominated, is no longer adequate.

One indicator of this is the spectacular growth of ICT spending in corporate America since the 1970s, which has outpaced other types of equipment expenditures, even after taking into account a growing workforce and the declining price of computer hardware and software. For instance, the nominal annual US corporate IT investment from 1970 to 2008 increased from about $5 billion to almost $350 billion. Overall IT investment rose exponentially in the United States, from about $100 per employee in 1970 to $3,000 in 2008. During the first decade of the twenty-first century, American industries became 5.5 times more ICT intensive than they had been in 1995.[7]

Interestingly, economists disagree about the extent to which this vast expenditure has led to productivity growth. Robert Gordon, for example, argues that the inventions of the period 1870–1900, including electricity, running water, and the motor car, resulted in far greater productivity gains than the Internet, the web, and e-commerce combined.[8] This issue is difficult to resolve, as it is well nigh impossible to disentangle productivity measures from the massive organizational changes wrought by information technology. As Manuel Castells stresses, new information technology is redefining workers, work processes, employment, and occupational structures, and the forms of this transformation have been the result of the "interaction between technological change, the institutional environment, and the evolution of relationships between capital and labor in each specific social context."[9]

A comparative overview of the impact of the *informational paradigm* on economic activity and levels of employment in different countries is provided by Castells, so I need not repeat that here.[10] The important point to note is that while the relationship between technology and the quantity and quality of jobs is complex, the general trend is toward an increasing divergence between work patterns and between workers. On the one hand, there has been a large rise in highly educated managerial, professional, and technical occupations and, on the other, a growth in low-end, unskilled, service jobs. In the Anglo American context, this has been described in terms of the dual growth of good MacJobs (such as professions in computing and information technology) and bad McJobs (work in fast food, retail, and personal services).[11] While high-end jobs are characterized by enhanced autonomy, low-end jobs are increasingly precarious (part time, temporary, and fixed term). The tendency for flexible working patterns cuts across these divisions but has very different consequences depending on the sector and type of job.

Overall, the rapid diffusion of ICTs into workplaces has fundamentally altered how work is performed. Indeed, the Pew Internet survey describes most working Americans as "networked workers" because they use all three basic tools of the information age: the Internet, e-mail, and the cell phone.[12] The trend is striking. In its first survey in March 2000, Pew found that 37 percent of full-time workers and 18 percent of part-time workers had Internet access at work. By mid-2011, this had grown to 76 percent of full-time workers and 52 percent of part time workers who use the Internet on the job. Sixty percent of workers use the Internet every day at work, while only a quarter say they never use computers. Almost all workers (87 percent) who have e-mail at work check their work e-mail at least daily. Half check their e-mails several times an hour, with just over one-third checking constantly. In addition, most workers have a mobile phone, and texting colleagues is common.

Since the survey was completed, the purchase of smartphones, tablets, and laptops has soared and people are spending more and more time online. The mean amount of active Internet use at work has doubled from 4.6 hours per week in 2001 to 9.2 hours in 2010.[13] This means that those who have Internet access at work are spending the equivalent of an entire workday per week online. Such estimates are notoriously difficult to make, as computers are on all day and are increasingly integrated into the work environment. It is also hard to get an accurate picture of what precisely the Internet is being used for. The *Networked Workers* survey found that, while at work, 22 percent of workers shop online, 15 percent watch videos, and 10 percent use online social or professional networking sites, while 3 percent play games. News consumption has also moved online and into the workplace as has pornography.[14] The workplace use of ICTs is by no means confined to work purposes, and for some provides access to forms of relaxation at work.

What the data clearly show is that the frequency and type of technology usage at work varies significantly with different kinds of jobs. Nearly three out of four professionals and managers use the Internet at work, either constantly or several times a day. About half of clerical, office, and sales workers also use the Internet at work at least several times a day. Daily Internet use is much lower in other job categories. These patterns hold for cell phone use with one central difference. Workers in skilled trades (for example, electricians) have relatively high rates of work-only cell phone use, compared to the proportion of skilled trades people that report daily Internet use.

The enormous variation in ICT usage across different occupations and industries, even within North America and Europe, makes it impossible to generalize. It is not my aim to present an overview of all the different worlds of work with ICTs. Rather, I want to focus on the central claims that are made about the impact of ICTs on the working patterns of highly networked workers: the intensification of work and the extension of work beyond the workplace (discussed in chapter 6).

Large-scale surveys like Pew typically report that such employees express mixed views about the impact of technology on their work lives. On the one hand, they cite the benefits of increased connectivity and flexibility that the Internet and all their various gadgets afford them at work. On the other hand, many say these tools have added stress and new demands to their lives. Curiously, few workers feel as though e-mail alone has increased the total amount of time they spend working.[15] In fact, most studies measure work "demands," not work hours, and therefore the connection between ICTs and longer work hours has not been established.[16]

Perhaps the time pressure we associate with ICTs, then, has as much to do with changes in the quality of working time, that is, an increase in the tempo or intensity of work, requiring us to work harder or faster. There are lots of ways we could think about this aspect of time pressure. The most interesting angle, in my view, is to explore three related aspects of work intensification: the pace of work, interruptions, and multitasking.

ICTs and the Pace of Work: Too Much Too Quickly?

Few American researchers have explored the impact of digital technology on work pacing.[17] However, for many years the British economist Francis Green has argued that the diffusion of ICTs is correlated with an increase in work effort, that is, a rise in "the proportion of effective labor performed for each hour of work."[18]

This is evident in the finding of a British employment survey that the number of workers who report "frequently working at very high speeds or to tight deadlines" has risen to record highs.[19] This is especially the case in workplaces where ICTs have recently been introduced. The report notes that the drivers of this intensification of work effort include management's ability to monitor the flow of work, employees' engagement with the new technology, and the heightened competitiveness brought on by the severity of the recession. While it is difficult to separate out the interconnected causes of rising effort levels, it is clear that ICTs increase management's ability to track workflows and improve coordination and monitoring of employees' performance, especially when staff are away from the office. Using ICTs, documents can be sent and received on the move while the flow of work is observed and maintained. Therefore, schedules can be altered at short notice, improving the microcoordination of tasks. Such flexibility encourages discretion, allowing an identification with project tasks rather than conformity with regulations, resulting in greater employee engagement. Green thus believes that the intensification of work is the result of managerial practices facilitated by technology and is not some pure effect of the devices themselves.[20]

The extent to which employees have discretion and control over the pace with which they accomplish tasks is key to feelings of time pressure. My own Australian study on the impact of mobile phones at work concurs with Green that ICTs have intensified work effort.[21] Especially for the men in my sample, using a mobile phone frequently while at work is significantly associated with both time pressure and other subjective indicators of work intensification. Frequently working under stress, working at speed to tight deadlines and time pressure are significantly correlated. However, does this mean that the mobile phone causes greater time pressure and stress? Or do those with stressful jobs who are under more intense time pressure use their mobile phones more often?

I concluded that it is not possible to determine the causal direction of the association. Workers are likely to experience more work, at an intense pace, under greater time pressure with more stress and heavier use of the mobile phone, as a single package. In other words, we need to be wary of making facile assumptions about the relationship between acceleration of the pace of work and ICTs.

Nowhere is this more the case than in discussions of information overload. Over the last decade, the increasing volume of e-mail has been talked about as a major source of stress. Countless management gurus advise us on how to take control and overcome the tyranny of our addiction to constantly checking e-mails. The legendary tale has Intel's chief executive Paul Otellini instituting "no e-mail days" after criticizing his employees "who sit two cubicles apart sending an e-mail rather than get up and talk."[22] He did this out of conviction that more direct communication would boost productivity. Designed to be a speedy and time-saving technology, e-mail appears to have bogged us down in endless time-consuming exchanges. So is e-mail the main culprit?

Few studies directly measure if, and how, e-mail overload contributes to the stress people experience. Typically, researchers infer stress by arguing that e-mail increases working hours. So the findings of an exceptional case study of knowledge workers in a high-technology firm are particularly striking.[23] That study found that people blame e-mail for the stress they experience, regardless of the amount of time they work and the fact that other communication activities also exacerbate their workload and the stress they feel. In other words, e-mail is not just a source but also a cultural symbol of overload people experience in their lives. Moreover, the authors argue, "by serving as a symbol, e-mail distracted people from recognizing other sources of overload in their work lives." Let me elaborate.

The study set out to untangle the contradictory findings that e-mail is a growing source of stress while, at the same time, it enables people to gain control over their work and reduce overload, thereby reducing stress. So why do workers attribute their stress exclusively to e-mail?

The answer lies in a sociomaterial approach that recognizes social processes "as important as—if not more important than—a technology's material properties for shaping its use and consequences."[24] Accordingly, and in close affinity with my own method, the authors examined the following: how e-mail's material properties interact with the anxieties that e-mail evokes, the norms that govern its use, and the temporal distribution of communicative acts over the course of a day.

Nearly half of the knowledge workers (engineers, technical writers, marketing personnel) associated e-mail with a loss of control, in that they feared falling behind and missing important information. Using e-mail eased their anxieties and allowed them to feel as if they were in control. Interestingly, the more time spent on e-mail, the more overloaded they felt, but the more messages they dealt with, the more they felt they could cope.

However, the speed of response was critical. Although e-mail's asynchrony, or time-shifting property, should allow recipients to respond at a time convenient for them, in fact most employees use e-mail in ways that promote rapid response. This reflects a shared *norm of responsiveness*. Those who answered e-mail quickly were revered, while other employees described sanctioning processes that were enforced when coworkers did not adhere to this organizational norm.

Finally, the authors argue that the flow of daily communications also matters. When people spend significant parts of the day in other activities, e-mail builds up and becomes figural just as the workday is about to draw to a close. As a result, rather than complaining about how much time teleconferences and meetings consumed, interviewees focused on the inbox as the salient source of overload. E-mail provides a "culturally sanctioned rhetoric of complaint about overload as well as a tangible ritual for regaining control: to cope with overload, trim your inbox."[25]

It is the sociomaterial distinctiveness of e-mail that explains why it is a symbol of overload. Because e-mail functions to temporally decouple responses from messages sent, time away from continuous processing results in a build-up in the inbox. In order to maintain control, given the obligation to answer quickly, workers extended their workday. Their experience of meetings, teleconferences, and phone calls was vastly different. Because these activities are synchronous, requiring copresence for communication, they left no material reminder of unaccomplished work.

In sum, e-mail is singled out as a symbol of stress because of this entanglement of material, social, and quasimaterial factors surrounding its use. The authors conclude that the focus on e-mail masks the real causes of overload: new demands of work that crowd days and create unrealistic expectations about response time.

This study rightly interprets the temporal significance of e-mail as not simply due to the speed of data transmission, which conventional accounts would have us believe. Focusing on only one type of ICT, however, has its limitations. There is mounting evidence that people are more and more using a variety of simultaneous modes, being perpetually available via *multimodal connectivity*. Therefore, it is important to investigate the extent to which the resulting constant interruptions that fragment working time lead to harriedness.

Rethinking Interruptions at Work

Sylvia Ann Hewlett is known for her articles in places like *Harvard Business Review* on what she terms "extreme jobs,"—seventy-plus hour workweeks.[26] She identifies six main "stressors" that cause exhaustion and burnout among people who work these long hours: rigidity combined with unpredictability, fast pace and tight deadlines, availability 24/7, constant travel, and work-related events outside regular work hours. But the sixth and chief stressor is the number of frustrating interruptions to the working day because of "canny communication devices." With several screens on a worker's desk, she says, an average of three minutes' attention is given to any single task without interruption. This scrambling and jarring of attention kills any hope of serenity, and, at the end of the day, there's usually a commute back to a domestic life of multitasking home and children.

There is a growing literature on the subject of interruptions by management and information systems scholars. This writing views interruptions as disruptive and negative, raising concerns about lowered productivity and time use. Defined as "a synchronous interaction which was not initiated by the subject, was unscheduled and resulted in the recipient discontinuing their current activity," interruptions are treated as single, isolated events that need to be managed.[27] Studies in this vein typically quantify them in order to evaluate the cognitive cost of interruptions and fragmented work. Research is then directed toward improving the design of technological interfaces, such as e-mail filtering software, in order to assist employees.

There are a number of problems with this mechanistic approach. First, there is an underlying assumption that interruptions divert employees' attention away from their "real" work. Accordingly, interruptions should be either minimized or, at the very least, the impact of them minimized. Second, employees are largely viewed as passive in the face of these interruptions. Their only response is to attend to the call for their attention. In sum, ICTs are viewed as exogenous, and distinct from the normal day-to-day operation of organizations.

By contrast, I treat technology as integral to doing work: organizational practices are mutually shaped with machines.[28] Such human-machine interactions crucially depend on the locally contingent meanings that people attribute to them and, therefore, the way workers interpret these "interruptions" from multimodal media is key. Knowledge workers now inhabit an environment where ICTs are ubiquitous, presenting simultaneous, multiple and ever-present calls on their attention. These mediated interactions can no longer be only framed as sources of constant interruptions that fragment and squeeze time.

This became apparent to me in the course of my own study, which is therefore worth expounding at some length. The research was carried out over a number of years, and it convinced me that the nature of knowledge work itself is changing as workers navigate the media-rich ecology of contemporary office life. As there is still a dearth of empirical research on these issues, I will report on my findings here.

Table 1. Frequency and duration of work episodes

Episode duration, in minutes	Mean number of episodes per day	Share of episodes, %	Mean duration of each episode, in minutes	Mean total time spent per day, in minutes
10 or less	79.1	90.4	2.9	229.7
11 to 30	6.3	7.2	16.9	106.8
31 to 60	1.7	1.9	44.1	73.5
Greater than 60	0.4	0.5	86.1	38.3
All episodes	87.6	100.0	5.1	448.3*

* This is equal to the mean total work time of approximately 7.5 hours.

Wanting to explore how work is being organized in and with the full range of ICTs, I collected detailed data on a group of knowledge workers in a multinational telecommunications company.[29] The fieldwork took place in the Sydney headquarters, a purpose-built campus colocated with a major university. Designed to facilitate informal interactions between staff of diverse roles and seniority, offices are open plan with partitions delineating workspaces of one to four employees. When standing in their work area, an employee typically has between twenty and a hundred workers in view.

The study involved shadowing participants at work and recording all episodes of work activity: their nature, location, duration, and the type of technology used. Interruptions were defined as the reason why the person changed their work episode, whether in response to an e-mail alert, phone call, or someone visiting a worker's desk. Changes in work episodes that were not directly attributable to an outside stimulus are referred to as self-initiated.

So how is multimodal connectivity affecting temporal practices at work? Let us begin by examining the extent to which workdays are fragmented and the nature of the activities carried out in these work episodes. A work episode refers to any of the various activities in which a worker participates: taking part in meetings, talking on the telephone, attending to e-mails on their computer, preparing a Microsoft Excel spreadsheet or getting a cup of coffee.

As table 1 shows, workdays are made up of a large number of work episodes—an average of eighty-eight per day—and the vast majority of

Table 2. Frequency and duration of communication and other activities

Communication activities	Mean number of episodes per day	Mean duration of each episode, in minutes	Mean total time spent per day, in minutes
Direct, nonmediated	**23.2**	**6.8**	**158.7**
Face-to-face	20.4	4.4	90.7
Meetings	2.8	24.5	67.9
Mediated	**37.1**	**4.7**	**172.8**
Voice call via landline telephone	12.9	5.1	66.0
Voice call via mobile device	4.9	3.3	16.1
E-mail via computer	16.9	5.1	85.8
E-mail via mobile device	2.3	2.1	4.9
Other activities			
E.g., report writing, tidying desk	**27.3**	**4.3**	**116.8**

these (90 percent) last for ten minutes or less. The average duration of work episodes is just under three minutes. Indeed, half of the workday is spent on work activities that last for ten minutes or less. The lack of sustained time knowledge workers spend on a particular work episode fits with common perceptions of fast-paced or intense work.

But what role do digital technologies play in these frequent changes in activity? Before attributing any role to ICTs for these interrupted work patterns, let us take a closer look at the nature of the work episodes and how ICTs are actually being used.

To my surprise, workers spend most of their time at work (about five and a half hours) engaged in communication activities (see table 2). However, when divided into direct, face-to-face communications and mediated communications (e-mail, telephone calls), over half of it is technologically mediated. While face-to face communication is still very important, ICTs are now integral to the way knowledge workers carry out their work roles. Strikingly, all forms of mediated communication are short, on average lasting less than five minutes.

ICT usage seems synonymous with short work episodes that fragment workers' workdays. The widespread perception is that technologies are the *cause* of these short work episodes, as mediated communications interrupt work on a particular task. Contrary to my expectations, I found that these technologies are a minor reason why people frequently change work activities during a workday. By contrast, face-to-face interruptions from other colleagues in the open-plan office take place an average of twelve times per day. But by far the greatest initiator of distinct work episodes are workers themselves—being the reason for change an average of sixty-five times per workday.[30]

What can one conclude from this? One interpretation is that workers are exerting control over their media-saturated environment and, indeed, taking advantage of ICTs to manage their availability. This runs counter to the common perception that ICTs dictate people's workdays. But in order to get a more nuanced picture, we need to situate workers' relationships with ICTs *within* the everyday work practices of their organization.

Fragmented workdays are not a new phenomenon. They were identified as a key aspect of managerial work in the 1970s.[31] What is new is the extent to which attending to mediated communications has become a normal and essential part of work. Knowledge work today is largely organized through mediated communication modes, as information and instructions are delivered via the various devices and applications in workers' communications repertoire. Given this, the common view that workers experience mediated communications as interruptions to their real work is a misnomer.

Indeed, the participants in my study do not perceive incoming mediated communication as a negative distraction. In many cases it was positive in the sense that the communications informed workers of their tasks for the day and of the progress of matters with which they dealt. When asked, both senior and middle managers typically expressed the view that interruptions actually constituted part of the workflow and were regarded as normal business rather than as disruptive. Rather than being distractions, then, interactions via technologies are an essential part of knowledge work.

Moreover, the capacity of ICTs to store messages electronically in their material memory means that constant connectivity does not inevitably mean constant interruptions. The asynchronous, time-shifting properties of ICTs allow workers greater input into how they respond to contemporaneous messages. They engage in an ongoing process of judging an appropriate level of availability, reviewing incoming information, assessing the priority of this information with respect to other tasks they are dealing with, and reordering and rescheduling their work activities.

Guiding this process is a complex interplay between the material affordances of the devices, workplace norms, and the decisions workers make as they perform their respective organizational roles.

In the organization I studied, for example, a hierarchy had developed that reflected the level of importance ascribed to different modes of communication. E-mail indicated that the communication was not urgent, while a landline telephone call might be more important. However, the use of a mobile device—be it for text messaging or voice communication—signified that the matter was very important and that the recipient needed to be available for that communication. Managers said that if a prompt response was required, then they would expect to receive a text or mobile phone call. Similarly, a call on a landline would not be regarded with the same urgency as a call on the mobile, in which the caller is identified on screen.

Depending on the mode of transmission, then, some electronic alerts are responded to immediately, while others are ignored, at least temporarily. How much control workers exert in the way they respond to these communications is critical and this, in turn, depends on the corporate culture. Interviewees reported that the high value placed on keeping on top of ever-changing priorities in their dynamic company meant that new work issues arise frequently and need to be dealt with quickly. This often requires the reordering and rescheduling of work tasks. Rather than simply responding to new or stored messages, workers stay connected in order to monitor work developments. The ways in which they interpret and behave with ICTs, therefore, can only be understood in their concrete and practical application in the organizational context.

I have reported at length on my study for the light it sheds on the relationship between interruptions, multimodal connectivity, and fragmented, pressured time. Although workdays largely consist of short work episodes, many involving mediated communication, this does not mean that the rhythm of work is set by the pace of machines. ICT-based practices do not ineluctably promote intensification by generating interruptions. My finding, that knowledge workers initiate the majority of changes in their work activity themselves, indicates that they are actively negotiating a whole new communication repertoire and deploying them as everyday tools.

In other words, the relationship between technology, pace, and connection is much more rich and complex than previous studies have shown. ICTs actually have an ambiguous relationship to the pace of work flow, both impeding and propelling, depending on a variety of local, human/machine contingencies.

I was intrigued to read a study of heavy BlackBerry users that highlights what the authors term the "autonomy paradox."[32] The corporate lawyers, venture capitalists, and investment bankers interviewed overwhelmingly regard "anytime, anywhere" mobile e-mail as enhancing their flexibility, control, and competence as professional workers. Yet, this same pattern of compulsive connectivity also heightens expectations of availability and responsiveness, reducing their personal downtime and increasing stress. The authors explain this paradox—that mobile e-mail both increases and diminishes professionals' autonomy—in terms of the unintended consequences of collective use. While individuals derive substantial benefits from their mobiles, monitoring and controlling the timing and location of work, the shared practice of continual connectivity fosters expectations of accessibility, escalating engagement with work at all hours of the day and night.[33]

Consistent with my argument, the authors show how, over time, the collective trajectory of ICT use shifts norms of how work (in this case, professional work) is and should be performed: redefining "what it means and what it takes to be an effective knowledge professional in an era of ubiquitous, always-on, mobile technologies."[34] In contexts where perpetual availability via multiple gadgets is a significant part of professional self-esteem, it may well be experienced as a positive attribute.

Judging how digital technologies will be incorporated into work processes merits continuous inquiry, especially as their very meaning changes over time as people innovate and modify them in their usage. For example, e-mail has already become a less urgent, asynchronous way of communicating, compared to when it was first adopted. Shifting between synchronous and asynchronous modalities, people are developing multidimensional time practices, creating new rhythms of work. The traditional concept of work intensification cannot fully capture this dynamic coconstruction of ICTs and the temporalities of work.

CHAPTER FOUR

Multitasking

ICT usage is deeply woven into the texture of workplace culture. One related feature of this new work pattern, which I did not explore in my own study, is the increasing likelihood of performing activities contemporaneously. Multitasking is the final aspect of work intensification that I want to consider here.

As I outlined in the previous chapter, time pressure can take a variety of forms, depending on which aspect of temporality is being squeezed. In addition to the volume of time available and the problem of temporal coordination, I introduced the concept of *temporal density*. This describes the experiences of juggling and multitasking, that is, the distribution of certain practices within temporal rhythms that create a sense of intensity in their conduct. I discussed temporal density in the context of the time-use literature that charts how working parents, especially mothers, juggle their paid work and domestic tasks. We saw significant growth in parents combining leisure with child care and I speculated that this might affect the distinctive quality of leisure time.

Research on multitasking with digital technologies is in its infancy. High levels of multitasking at work, however, are directly linked to ICT use.[35] Multitasking is seen as an efficient way of dealing with interruptions, integrating them into workflow and thus saving time. For example, using e-mail while teleconferencing or attending a lecture has become commonplace. It is justified on the basis that people are able to pay attention to any number of tasks simultaneously.

But several studies indicate that multitasking may actually impair performance. Ironically, "heavy" media multitaskers are more susceptible to distractions and perform worse on cognitive tasks than "light" multitaskers.[36] So increasing one's amount of ICT-based multitasking does not improve ability to master it. In general, the literature on multitasking points to negative outcomes: reduced cognition or performance (even among "digital natives") or rising work-related strain. Psychologists confirm that humans are incapable of giving their full attention to two tasks simultaneously.[37] What people actually do is switch their attention from one task or platform to the next, and such task switching leads to a host of issues, including attention difficulties, poor decision making, and information overload.

Nevertheless, it may be that all multitasking is not of the same kind, as all tasks do not require the same level of attention. Certainly Lee Rainie and Barry Wellman are sanguine about this, remarking that thriving networked individuals have "*multitasking literacy*: the ability to do several things (almost) at once."[38] They deal with multiple inputs from family, friends, work and institutions, via multiple devices, "without fuss." The authors dismiss skeptics by noting that driving a car requires precisely such skills. Clearly they are not worried by the reports that talking on a mobile phone while driving is associated with more accidents!

The notion of multitasking brings to mind the well-known distinction between *monochronic* time, in which events are scheduled as discrete, separate items, and *polychronic* time, in which several things are done simultaneously. This distinction has been much discussed by anthropologists in exploring cultural differences in perceptions of time. Edward Hall argues that the dominant time culture of the United States and Northern Europe is monochronic, and that the contemporary workplace is organized according to this "one thing at a time" assumption. Polychronic time, however, takes over in the home, "particularly the most traditional home in which women are the core around which everything revolves." Hall even argues that, like oil and water, the two systems do not mix, and that at the "pre-conscious level" monochronic time is male time and "polychronic time" is female time, and women therefore find it hard to conform to the alien time culture of work.[39]

The gender dualism implied by such dichotomies has been the subject of much feminist critique. Even cognitive psychologists who seek out such differences have failed to find conclusive evidence that a gender-based *predisposition* to polychronicity exists at all.[40] It would be surprising if it did, as these same psychologists confirm that (male-dominated) managerial and executive work are quintessentially polychronic, compared with nonmanagerial work. Dichotomies aside, the idea that many temporalities coexist and that the human experience of time is not uniform is key to my thesis. One only has to think of the linear temporal logic of capitalist production exemplified by scientific management, compared to the relational time involved in the physical and emotional care of others.

Moreover, we should be wary of assuming that any one particular time orientation is superior to another. On the basis of an extensive review of the effectiveness of polychronicity in organizations, Allen Bluedorn concludes that stress-related findings are mixed and that sometimes monochronic behavior patterns are best, sometimes polychronic.[41] Although these organizational studies do not specifically examine the effects of ICTs, they point to the contingent nature of these temporal strategies for individuals and groups.

Bluedorn thus questions the common notion that polychronic life strategies lead to negative health outcomes. Many studies show that the density of social networks is related to positive health outcomes. The introduction of ICTs that facilitate this kind of relational multitasking is likely to be positive for well-being. For example, Noelle Chesley documents a positive association between personal forms of ICT use and higher ratings of "feeling effective at work."[42] This is particularly the case with cell phones, and it may be that people already in demanding jobs use cell phones to coordinate and reschedule personal tasks to accommodate work, a subject I will explore in the following chapter.

What is clear from these studies is that the personal context in which multitasking takes place is of the utmost importance. Studies of multitasking that do not involve ICTs, which are much more common, confirm this. Multitasking has become pervasive because of the increased demands at work and at home, and mothers, more than fathers, generally experience multitasking as negative and stressful. Shira Offer and Barbara Schneider found that for both mothers and fathers in dual-earner families, multitasking can be a positive experience when in the company of other family members, like a spouse or child.[43] By contrast, for both, multitasking at work, although associated with an increased sense of productivity, is generally perceived as a negative experience.

Multitasking as a time-management strategy is, then, a common feature of dual-earner family members' busy lives. However its usefulness as a concept is hard to gauge. At the very least, it should be distinguished from polychronicity, as the term *multitasking* implies that performing multiple activities increases productivity, that is a form of work intensification. But the idea does focus attention on the quality of temporal density, and this, rather than the amount of time, per se, is consequential for perceptions of time pressure.

Neither interruptions nor multitasking are new and both long predate the introduction of digital technologies. It is too easy to presume that extensive use of ICTs is the primary cause of time pressure rather than a symptom of structural changes in the conditions of work. The lesson of the above studies is precisely that whether multitasking contributes to or alleviates time pressure depends on the broader social context in which it occurs. And so I would like to conclude on this theme.

Conclusion

The dramatic restructuring of workplaces in the face of economic forces and heightened global competition has meant that there is often more work to do, with fewer people to do it. Networked workers' jobs have also become more complex and demanding as they are continually required to update their skills to keep up with the latest technology. ICTs increase the speed and ease with which information can be gathered, processed, analyzed, and shared, fostering a higher volume of mediated communications. In combination with a cultural norm of speedy responsiveness, the pace of work quickens. At the same time, as new technologies become routinized, we are seeing the emergence of radically different temporal rhythms and modes of working.

Working practices are always inscribed in the dense materiality of the sociotechnical infrastructure. Today information networks "shape not only the work routines, but also the ways people look at practices, consider them 'natural' and give them their overarching character of necessity. Infrastructure becomes an essential factor shaping the taken-for-grantedness of organizational practices."[44] This does not mean that ICTs have effects independent of human actors. Information technologies that augment management control and extend the reach of surveillance also reconfigure the ways in which work is performed, and generate new economic and social arrangements. Indeed, digital technologies are uniquely malleable and open to modification and interpretation by users. An STS lens directs our attention to the complexities and contingencies of technological developments. In doing so, it immediately casts doubt on the determinist view that ICTs, per se, are driving the intensification of work.

That networked workers' everyday experience of these technologies is riddled with contradictions is hardly surprising. On the one hand, ICTs are a source of personal autonomy and flexibility, enhancing control over one's availability. On the other hand, in downsized, high-pressure workplaces, these same tools add stress and escalate collective expectations of constant communication and engagement with work. Paradoxically, the more intense pressure people feel, the more they deploy digital devices in order to save time.

The profound ambivalence that people feel is all the more understandable given the fundamental transformation that we are witnessing in the nature of capitalism, work, and the labor market. Sociologists like Ulrich Beck and Anthony Giddens have written extensively about the accelerated pace of social change in late modern societies. They emphasize the sharp decline in both the power and legitimacy of authoritative norms and institutions and the increasing uncertainty of all social arrangements. Both connect the sea change in how employees understand themselves and their work to these developments. However, it is Richard Sennett who most compellingly portrays the "short-term" temporal horizons that are the hallmark of the contemporary work experience.

In *The Corrosion of Character: The Personal Consequences of Work in the New Capitalism*, Sennett argues that the loss of the "long term" in the new capitalism is producing a crisis of character.[45] In the old industrial economy, the possibility of sustained, predictable, usually lifetime engagement with a workplace was the bedrock of people's identities—their "character." Work provided status, dignity, and the opportunity for self-development. With its principle of "no long term," however, the new, flexible capitalism has torn away the basis for the formation of sustaining identities, resulting in a growing malaise as people search in vain for a place to anchor a meaning for their lives. Modern work is fragmented, insecure, stressful, and unpredictable. Without the guarantees of lifetime employment and a steady-state career trajectory, workers have little incentive to be loyal to or trusting of their bosses, coworkers and corporate employers. Sennett diagnoses an increasingly lost workforce, unable to build the social relationships that lie at the heart of a fully developed, sustaining social identity.

Although this linear, predominantly male biography was never as standard as Sennett claims, it is true that the old contract between work and time has broken down.[46] The historical trend of salaried work and socialized production has stalled. Occupations that extend over the whole of a work life are rare, and job tenure is shorter than it was a generation ago. Sennett's account resonates strongly with those employed in professional and managerial jobs who have seen secure bureaucratic careers erode and the shift of market risks and responsibilities onto individuals themselves. Emblematic of this truncation of time spans is the spread of performance management systems, which audit individuals on current performance, without regard to a person's history of effort. Such seismic shifts in the wider culture inform our overall sense that time is telescoping.

This chapter has focused on how information and communication technologies affect time while people are at work. However, one of the main claims about the impact of ICTs is their propensity to extend work into times and spaces traditionally work-free. The Pew survey shows that networked workers use digital devices across social domains, including home, to accomplish work.[47] I will examine this claim, the so-called blurring of personal and work time, in detail in chapter 6. But first I want to consider the impact of technology on domestic work. If the time deficit is as much to do with the combined demands of work and family life as it is to do with technological change, then the time spent in housework is a critical factor in people's experience of harriedness.

SAMSUNG

Chapter Five

Doing Domestic Time

The washing machine has changed the world more than the Internet has.

HA-JOON CHANG, *23 Things They Don't Tell You about Capitalism*

Advances in technology have transformed the work that goes on in the household. Indeed, it is not too much to claim that an industrial revolution occurred in the home, that "the change from the laundry tub to the washing machine is no less profound than the change from the hand loom to the power loom."[1] As in other spheres, considerable optimism has attached to the possibility that technology may provide the solution to time-consuming drudgery.

Many economists and sociologists have argued that it is precisely the time-saving properties of domestic technologies that freed women to enter the labor force. For example, Ha-Joon Chang argues that household appliances, such as the washing machine, have been revolutionary in that they liberated women from doing tedious housework and led to the demise of the domestic servant.[2] While such arguments can tend toward technological determinism, they are refreshing in attributing significance to what are generally regarded as mundane household artifacts.

This chapter will explore the relationship between domestic technology and the time households allocate to domestic work (leaving ICTs largely to the next chapter). In particular, I will grapple with the conundrum that devices allegedly designed to save labor time fail to do so, and in some cases actually increase the time needed for the task.

Household technology has largely been invisible in academic discourse about the acceleration society, and this is not unrelated to the invisibility of housework. Yet, as we saw in chapter 3, a principal reason for the perception that life has become more rushed is the rise of dual-earner households, with work time squeezing the time available for housework and caring activities. Any discussion of household labor is always at the same time a discussion of gender roles and relationships. So gender differences in perceptions of time and time pressure will be a theme throughout this chapter.

I begin by providing some historical background to the mechanization of housework, and then consider contemporary arguments about the impact of technological innovation on the ways in which households assign time to housework. I will examine the reasons why "time-saving" domestic technologies, such as washing machines and microwaves, have been more successful in time shifting than they have been in reducing the amount of time spent in domestic work. Changed cultural expectations about parenting, as we have already noted, point to the peripheral effect of technology. Here, however, my focus is on the related issue of how domestic activities and artifacts themselves are intertwined with gender identities. The way in which men and women divide their time in the home, and the type of housework they do, impacts upon their differing feelings of harriedness. Finally, I examine the cultural imaginaries around the digital home of the future, showing how they reflect the mindset of their designers.

The Industrialization of the Home

What was the relationship between the technological developments in the economy and those in the home? To what extent did new technologies "industrialize" the home and transform domestic labor? Why, despite massive technological changes in the home—such as running water, gas and electric cookers, central heating, washing machines, and refrigerators—do studies show that household work in the industrialized countries still accounts for approximately half of the total working time?[3]

The conventional wisdom is that the forces of technological change and the growth of the market economy have progressively absorbed much of the household's role in production. The classic formulation of this position is to be found in Talcott Parsons and Robert Bales's functionalist sociology of the family.[4] They argue that industrialization removed many functions from the family system until all that remains is consumption. For Parsons and Bales, the wife-mother function is the primary socialization of children and the stabilization of the adult personality; it thus becomes mainly expressive or psychological, compared with the instrumental male world of "real" work. More generally, modern technology is seen as having either eliminated or made less arduous almost all women's former household work, thus freeing women to enter the labor force. To most commentators, the history of housework is the story of its elimination.

Although it is true that industrialization transformed households, the major changes in the pattern of household work during this period were not those that the traditional model predicts. Ruth Schwartz Cowan, in her celebrated study of the development of household technology between 1860 and 1960, argues exactly that.[5] For her, the view that the household has passed from being a unit of production to a unit of consumption, with the attendant assumption that women have nothing left to do at home, is grossly misleading. Rather, the processes by which the American home became industrialized were much more complex and heterogeneous than this.

Cowan provides the following explanations for the failure of the "industrial revolution in the home" to eliminate household tasks. Mechanization gave rise to a whole range of new tasks which, although not as physically demanding, were as time-consuming as the jobs they had replaced. The loss of servants meant that even middle-class housewives had to do all the housework themselves. Further, although domestic technology did raise the productivity of housework, it was accompanied by rising expectations of the housewife's role, which generated more domestic work for women. Finally, mechanization has only had a limited effect on housework because it has taken place within the context of the privatized, single-family household.

It is important to distinguish between different phases of industrialization that involved different technologies. Cowan characterizes twentieth-century technology as consisting of eight interlocking systems: food, clothing, health care, transportation, water, gas, electricity, and petroleum products. While some technological systems do fit the model of a shift from production to consumption, others do not.

Food, clothing, and health care systems do fit the "production to consumption" model. By the beginning of the twentieth century, the purchasing of processed foods and readymade clothes instead of home production was becoming common. Somewhat later, the health care system also moved into centralized institutions, and these trends continued with increasing momentum during the first half of the last century.

The transportation system and its relation to changing consumption patterns, however, exemplified the shift in the other direction. During the nineteenth century, household goods were often delivered, mail-order catalogs were widespread, and most people did not spend much time buying goods. With the advent of the motor car after the First World War, all this began to change. By 1930, the automobile had become the prime mode of transportation in the United States. Delivery services of all kinds began to disappear and the burden of providing transportation shifted from the seller to the buyer.[6] Meanwhile, women gradually replaced men as the drivers of transport, more and more business converted to the "self-service" concept, and households became increasingly dependent on housewives to provide the service. The time spent on shopping tasks expanded until today the average time spent is eight hours per week, the equivalent to an entire working day.[7]

In this way, households moved from the net consumption to the net production of transportation services, and housewives became the transporters of purchased goods rather than the receivers of them. The purchasing of goods provides a classic example of a task that is generally either ignored altogether or considered as 'not work', in spite of the time, energy and skill required, and its essential role in the national economy.

The last four technological systems—water, gas, electricity, and petroleum—also reorganized housework, yet their impact was ambiguous. On the one hand, they radically increased the productivity of housewives. "Modern technology enabled the American housewife of 1950 to produce singlehandedly what her counterpart of 1850 needed a staff of three or four to produce; a middle-class standard of health and cleanliness."[8] On the other hand, while eliminating much drudgery, modern labor-saving devices did not reduce the necessity for time-consuming labor. Thus there is no simple cause-and-effect relation between the mechanization of homes and changes in the volume and nature of household work.

Indeed, the disappearance of paid and unpaid servants (unmarried daughters, maiden aunts, grandparents, and children fall into the latter category) as household workers, and the imposition of the entire job on the housewife herself, was arguably the most significant change. The proportion of servants to households in America dropped from one servant to every fifteen households in 1900 to one to forty-two in 1950.[9] Most of this shrinkage took place during the 1920s. The disappearance of domestic servants stimulated the mechanization of homes, which in turn may have hastened the disappearance of servants.

This change in the structure of the household labor force was accompanied by a remodeled ideology of housewifery. In the early years of the twentieth century, the domestic science movement, the germ theory of disease and the idea of "scientific motherhood" led to new exacting standards of housework and child care.[10] As standards of personal and household cleanliness rose, the introduction of washing machines, for example, increased laundering to meet higher expectations of cleanliness. There was a major change in the importance attached to child rearing and mother's role. The average family had fewer children, but modern child-centered approaches to parenting involved spending much more time and effort.

Housework began to be represented as an expression of the housewife's affection for her family. The split between public and private meant that the home was expected to provide a haven from the alienated, stressful, technological order of the workplace and was expected to provide entertainment, emotional support, and sexual gratification. The burden of satisfying these needs fell on the housewife.

With home and housework acquiring heightened emotional significance, it became impossible to rationalize household production along the lines of industrial production. Rather than following Fordist principles of large-scale production systems and a cooperative application of labor, the dominance of single-family residences and the private ownership of correspondingly small-scale amenities is, as Cowan aptly puts it, a completely irrational use of technology and labor time within the home:

> Several million American women cook supper each night in several million separate homes over several million separate stoves—a specter which should be sufficient to drive any rational technocrat into the loony bin. . . . Out there in the land of household work there are small industrial plants which sit idle for the better part of every working day; there are expensive pieces of highly mechanized equipment which only get used once or twice a month; there are consumption units which weekly trundle out to their markets to buy 8 ounces of this nonperishable product and 12 ounces of that one.[11]

This was not inevitable. There were many examples of alternatives, such as British and American experiments in social housing in the 1930s, with communal restaurants and laundries.[12] However, these initiatives foundered and, consequently, domestic technology has been designed for use in privatized single-family households, requiring individuals to have a multiplicity of skills.

The relationship between domestic technology and the time spent on household work provides a good illustration of the general problem of technological determinism, where technology in itself is said to have resulted in social changes. The demise of domestic servants, changing standards of hygiene and child care, the ideology of the housewife's role, and the symbolic importance of the home reflect social as much as technological developments. Indeed, it is impossible to separate out changing housework practices from concurrent technological innovations.

Gender Specialization of Household Work

Domestic technology, then, has not directly reduced the time spent on housework. However, the way that men and women relate to technology may be a significant factor in determining how they spend time at home. This section will consider whether domestic innovations have had any effect on the gender specialization of household labor.

As I have already noted, a key factor fueling time pressure is the growth of the dual-earner family. Although this has been accompanied by a widespread belief in the value of sharing domestic duties in modern marriages, in reality, women are particularly harried because of their disproportionate load. Although men's contributions have increased substantially, women's labor still accounts for over two-thirds of the total time devoted to unpaid work.[13] While the sheer volume of work is bound to influence feelings of time pressure, so too will the kind of work performed. Here, therefore, I want to look at different types of housework that men and women undertake and what this means for their experience of time.

There has been a gradual convergence in gender patterns of domestic work between the 1960s and the 2000s. This emerges clearly from an extensive review of cross-national trends covering over twenty countries by Man Yee Kan, Oriel Sullivan, and Jonathan Gershuny.[14] They distinguish between different categories of domestic work as follows: routine housework (including daily routine types of housework such as cleaning, doing the laundry, and cooking), caring for family members (including care for children and adults), and nonroutine types of household work (such as shopping, gardening, and household repairs). Their main conclusion is that gender segregation between the different *types* of domestic work remains remarkably resilient.

While men are increasing their contribution in all categories of domestic work, they still spend comparatively little time overall on routine housework and much less time on child care. Men mainly do the less routine types of chores, such as DIY projects and shopping. Women still undertake the bulk of each type of domestic work, focusing particularly on routine housework (with cleaning, cooking, and laundry exhibiting the highest level of female specialization) and caring for others. Shopping and domestic travel times show a rising trend for both sexes and are less unequally divided, though women still do the largest part of these activities.

This traditional gendered specialization of household tasks is itself a major barrier to the equalizing of time use. Its persistence over the past forty years, despite women's increasing labor force participation, points to the continuing hold that gender-appropriate use of time has for feminine and masculine identities. Caring activities and routine cooking, cleaning, and laundry are strongly marked as female, whereas nonroutine chores such as DIY projects and outside work are male activities. Perhaps surprisingly, the authors note that these gender divisions do not seem very susceptible to change even in countries where gender ideologies are considered relatively nontraditional (such as the Scandinavian countries).[15]

The dominant characteristic of routine housework is that it is never complete. It is both less satisfying and more tiring than nonroutine household tasks. For women the home represents a sphere of work; for men it is a site of leisure, an escape from the world of paid work. Furthermore, a major problem with the research referred to above is that it does not recognize that the essence of housework is to combine many things, usually concurrently. This has a profound bearing on the interpretation of time spent in child care and the apparent growth of leisure. As I outlined in chapter 3, shopping accompanied by children lends leisure time a different quality than consumption as an individual leisure activity. While both male and female parents are devoting more time to child care, mothers typically remain much more emotionally and practically involved with their children than fathers. Moreover, they tend to have the management of—and responsibility for—the main elements of family life. Child care is perhaps the activity least amenable to technological solutions as time spent caring for others has a unique quality, a topic I will return to below.

Is Technology the Solution?

In the absence of a radical redistribution of the domestic workload within families, modern machinery holds out the promise of at least solving the problem of the routine types of housework, such as cooking, cleaning, and laundry. But what is the impact of so-called time-saving appliances, such as the microwave and dishwasher?

This question is harder to resolve than it may at first seem. Most researchers rely on the passage of the years as a proxy for ownership of appliances, since a higher proportion of contemporary households now own domestic appliances. In general, however, they do not have direct data about which households actually own or do not own particular domestic appliances. Nor do they provide information about the relationship between the appliances and time spent in the specific task for which they are designed. The Australian Time Use Survey is rare in providing such detailed information. Analyzing this data, I found that household technologies rarely reduce women's unpaid working time and even, paradoxically, produce some increases in domestic labor.[16]

Consistent with the cross-national findings above, Australian women are predominantly responsible for the routine housework of food preparation, cleaning, and laundry. Men spend more time on home, car and lawn maintenance, and outdoor jobs than on laundry and cleaning put together. Domestic appliances thus enter a domain heavily signified in terms of traditional sex roles.

Owning appliances makes little difference. Despite the microwave's capacity to cook food in a fraction of the time needed by conventional stoves, owning a microwave has no significant effect on the time-use patterns of women, even when the number of meals out is held constant. Nor does the deep freezer's ability to harvest economies of scale in meal production significantly reduce the average time that women devote either to meal preparation or to housework overall. Even dishwashers appear to have no effect on either the time women spend in food preparation and cleanup or in the daily hours they devote to housework. Owning a clothes dryer actually increased the time women spend doing laundry. However, some kitchen appliances, such as dishwashers and deep freezers, lead to reductions in men's housework time. Only a lawn mower or edge-trimmer increases the time men devote to the traditionally male tasks of lawn care.

These paradoxical effects, whereby inventions deliberately designed to save labor fail to do so, or even do the opposite, take some explaining.

We saw how, in the first half of the twentieth century, technological innovations spurred radical changes in the ways that people behaved. As appliances became widely diffused and accepted as necessary and normal, they altered the patterns and practices of life within households. Key here is the idea of rising standards in domestic production. The concept of rising standards implies a greater quantity or quality of domestic production—for example, more or better meals, cleaner clothes, and more attractive lawns. In other words, it may be that appliances are being used to increase output rather than to reduce the time spent on housework.

An indication of this is that one-third of all UK energy is now consumed in homes. The electricity consumption of domestic appliances has increased by 144 percent since 1970, largely due to the use of fridges and freezers, dishwashers, tumble dryers, washing machines, and other consumer electronics.[17] Energy use associated with cooking in the home is the only form of domestic energy consumption to have declined since 1970—a consequence of more people eating out. Water consumption within the home has increased by 70 percent over the last thirty years, and around 29 percent of domestic energy is used for heating water. Indeed, in 2010, domestic consumption was 32 percent of total UK final energy consumption, an increase of 31 percent since 1970. Kitchens and bathrooms in particular have become hot spots of resource consumption, reflecting changes in the way people wash, prepare meals, clean dishes and socialize.

Expectations of comfort, cleanliness, and convenience have altered radically over the past few generations, but these dramatic changes have largely gone unnoticed. Elizabeth Shove demonstrates this by taking three dimensions of "ordinary" life—the comfort of air-conditioning, the personal cleanliness of bathing and showering, and the convenience of contemporary domestic laundering—as probes into major changes in the fabric of daily life and social being over the last few decades. As these new domestic tools for living become a normal part of consumption, they escalate standards of comfort, shifting conventions, practices, and routines of consumption.

Let's take her example of bathing. Although "the technologies involved—the bath and the shower—have changed hardly at all over hundreds of years . . . patterns and logics of use are continually on the move."[18] How do we explain the increasing popularity of the shower, especially the power shower (which pumps out water at a higher rate), and the decline of traditional British bathing habits? Power showers far exceed the water and energy consumption of a twice- or thrice-weekly bath, however long the bather spends in the tub. In both the UK and the United States, the seven- to eight-minute daily shower is taken for granted as the norm for those who regularly shower.[19] This drive for personal cleanliness has intensified in a context where the shift from industrial to white collar service work and decreased air pollution has reduced the need for washing.

The constant availability of hot water and the heightened value of hygiene are prerequisites for the increasing frequency with which people wash. However, a key difference between bathing and showering is that the latter is associated with speed, immediacy, and convenience. The shower is not inherently quicker than a bath, but its potential to be used in this way fits our current preoccupation with saving time. The issue is not a literal lack of time itself. As I have argued, the experience of rushing is as much the result of the temporal disorganization of daily life. The proliferation of time-demanding social practices makes the achievement of collective routines harder to achieve. Even the once-weekly routine of a Sunday bath demands a form of temporal and sequential scheduling not required of privatized, fragmented episodes of washing. By contrast, showering can be slotted into narrow time frames—like those between waking in the morning and leaving for work—from which bathing is excluded. It is in this context that the ability to take a rapid shower acquires its appeal.

As such, the shower belongs to a set of domestic devices whose popularity has grown precisely because they promise to help people cope with the temporal challenges of modern life. For Shove, the new habit of showering can be understood only in relation to concerns about a time squeeze and its qualities as a technology of "convenience." The idea here is that appliances facilitate complex scheduling, delayed tasks as well as simultaneous activities. Washing machines, for example, allow clothes to be washed while you do other things, and with freezers, frozen food can be bought and used at a later time. As I noted in chapter 3, convenience food can be seen in these terms, as a hypermodern form of convenience that is directed toward time shifting rather than time compression.[20]

This time-shifting property of convenience technologies is harnessed during periods of intense busyness in order to generate and protect pockets of quality time.[21] Certain times of the day, for example, mornings and meal times, are more frantic than others. This is when the density of activity takes place and when time pressure is most acute. Domestic devices are used to preserve and manage distinctions between "rush" and "calm." However, according to Shove, convenience devices enter a feedback loop whereby they paradoxically increase problems of scheduling, exacerbating the sense of harriedness: "more gadgets generate more rush."[22] While the idea of convenience as legitimating and sustaining specific forms of consumption is illuminating, in my view it understates the myriad processes through which users interpret and appropriate machines. People are extremely adept at deploying suitable devices in their endeavor to make time for valued interactions. This is not only the case for domestic appliances but, as we shall see in the next chapter, also applies to ICTs.

Household Appliances: An Afterthought?

Thus far I have emphasized how new tools for living escalate standards and transform domestic practices. However, we rarely reflect on the material form of the household equipment that has become available to us, and why it is so. I have argued that the predominance of the single-family household has profoundly structured the design and configuration of our appliances. Little attention has been given to the innovation, development, and diffusion processes of specific technologies themselves.

Yet if there is one lesson that an STS perspective teaches us, it is that domestic artifacts, like other technologies, are both socially constructed and society shaping. Rather than speeding up existing activities, often as not they change the very nature and meaning of tasks and introduce novel practices. Given that much domestic technology has its origins in very different spheres, rather than being specifically designed to save time in the household, it is not surprising that its impact on domestic labor has been mixed. Indeed, the temporal landscape of the factory that informs the design of these technologies necessarily carries over into the home. Technologies that emerge as "transfers" from the production process in the formal economy to those in the informal domestic economy are unlikely to be attuned to the complex timescapes of the home.

Typically, new products are initially too expensive for application to household activities; they are employed on a large scale by industry only, until continued innovation and economies of scale allow substantial reduction in costs or adaptation of technologies to household circumstances. Many domestic appliances were initially developed for commercial, industrial, and even military purposes and only later, as manufacturers sought to expand their markets, were they adapted for home use. Gas and electricity were available for industrial purposes and municipal lighting long before they were adapted for domestic use. The automatic washing machine, the vacuum cleaner, and the refrigerator had wide commercial application before being scaled down for use in the home.

Electric ranges were used in naval and commercial ships before being introduced to the domestic market. Microwave ovens are a direct descendant of military radar technology and were developed for food preparation in submarines by the US Navy. They were first introduced to airlines, institutions, and commercial premises before manufacturers turned their eyes to the domestic realm.

Despite the lucrative market that it represents, the household is not usually the first area of application that is considered when new technologies are being developed. For this reason, domestic appliances are not always appropriate to the household work that they are supposed to perform. Nor are they necessarily the implements that would have been developed if the main user had been considered first or, indeed, if she had had control of the processes of innovation.

It is no accident that most domestic technology originates from the commercial sector, nor that much of the equipment that ends up in the home is somewhat ineffectual. As an industrial designer I interviewed put it, "Why invest heavily in the design of domestic technology when there is no measure of productivity for housework as there is for industrial work?" Commercial kitchens, for example, are simple and functional in design, much less cluttered with complicated gadgets and elaborate fittings than most home kitchens. Reliability is at a premium for commercial purchasers, who are concerned to minimize their running costs both in terms of breakdowns and labor time. Given that labor in the home is unpaid, the same economic considerations do not operate. Therefore, when producing for the home market, manufacturers concentrated on cutting the costs of manufacturing techniques to enable them to sell reasonably cheap products. Much of the design effort is put into making appliances look attractive or impressively high-tech in the showroom—for example, giving them an ornamental array of buttons and flashing lights. Far from being designed to accomplish a specific task, some appliances are designed expressly for sale as moderately priced gifts from husband to wife and in fact are rarely used.

Let us look in more detail at the microwave, as it is widely regarded as an iconic time-saving device. As we have just noted, microwave ovens were initially developed for use in submarines. When manufacturers first turned their eyes to the domestic market, they conceived of the microwave as a device to reheat prepared food for use by men, especially single men. As a result, microwaves were marketed as "brown goods" and sold next to hi-fi equipment, televisions, and video recorders—goods for leisure and entertainment. Even their color signified a gendered conception of household functions and consequently a gendered conception of potential purchasers. As it transpired, this attempt to frame demand was unsuccessful and subsequently the product was reconstituted as a simple and serviceable "white good" for use by housewives to cook.

In fact, women users appropriated this device in ways that were not foreseen by the engineers who designed it. Mapping the evolution of the microwave oven, Cynthia Cockburn and Susan Ormrod demonstrate how consumers can modify the meanings and values of technologies in the practices of everyday life. These cultural meanings, in turn, enter the design and production of goods themselves. Indeed, the authors conceive of technologies as in a continuous process of negotiation, as we "domesticate" or make new technologies our own. The finished form of the microwave, which redefined the gendered character of the user, meant that the microwave literally shifted its place in the department store. It now sits alongside washing machines, fridges, and freezers as a humdrum domestic appliance.

The making of the microwave, then, is as much a story about the transformation of a quintessentially human activity, cooking, as it is about a technical invention that saves time. It is perhaps the perfect hypermodern convenience device as it solves problems of meal coordination and synchronization. However, in order to grasp its full significance, we need to move beyond the focus on a single device.

Domestic technologies are adopted as part of larger organizational and technical infrastructures. Microwaves, for example, rely on freezers and are largely used as a defrosting machine. They also depend on a complex food supply chain involving an international workforce largely invisible to the purchaser. Convenience meals still require shopping in a context where car dependency and urban growth means traversing long distances to the supermarket. While the market for convenience foods has increased exponentially, so too has the practice of eating out, even though it requires temporal and spatial coordination. Today, about half of the money used to buy food in the United States is spent in restaurants.[23] Fast foods restaurants, such as McDonald's, are the main beneficiaries of this trend, and it is notable that they market themselves as "fast." The fast food industry exemplifies the way in which the capacity of more affluent households to buy time through utilizing consumer services is made possible by the supply of cheap, often migrant, labor.

Is Outsourcing the Solution?

The limits of even hypermodern technologies of convenience to compress time is reflected in the massive shift in the balance between privatized household work and the provision of market-based domestic services. The contracting out of housework is a growing trend in affluent societies, especially among middle-class households. The feminization of the labor force has greatly increased demand for the types of services that housewives traditionally perform, such as cleaning, cooking, and child care. Indeed, there is a direct link between married women's earnings and the time they spend performing routine household chores (cooking, washing the dishes, house cleaning, and laundry). As women who are employed full time earn more, they reduce the time they spend performing routine housework, "defray[ing] their housework time by using their earnings to purchase market substitutes, or services, for domestic labor."[24]

This phenomenon has received relatively little attention, largely because sociological research on housework has concentrated on the sexual division of domestic tasks. Feminist researchers have had to confront inequalities between women and address the contradictory interests that women have as employers and employees. After all, it is largely women who hire and fire other women to do housework and substitute mothering, as paid domestic servicing usually involves labor substitution for men in the household.

While the rich have always hired domestic servants, there is today a mass importation of caring labor from the third world. Many immigrant women leave their own children in order to work as nannies, cooks, and cleaners for families in the first world. From Hong Kong to the United States, for example, Filipina domestic workers are employed for low pay and long hours. Indeed, countries such as the Philippines have become economically dependent on the remittances women domestic workers send home. What is distinctive about this form of labor is that it straddles the boundaries of public and private and of free and unfree labor. In the case of live-in domestics, for example, what the employer is buying is not just the worker's labor power but their *personhood*.[25] This increasingly internationalized trade in emotional labor, "the global care chain" is based on intersecting hierarchies of sex, class, age, race, and nation.

Many domestic workers are employed not because people do not have the time to do their own domestic work but because they want to avoid doing the chores themselves and therefore gain the time for extra leisure. In other words, the cash-rich, time-rich—and not just the cash-rich, time-poor—also employ domestic workers. Those who do the work are often poor women with heavy child care responsibilities, hardly "time rich" themselves.

While hiring domestic labor to perform routine household tasks raises normative issues about the boundaries of self-maintenance, the outsourcing of care for family members (whether for children and adults) is far more controversial. Time may be squeezed, but both men and women have increased their participation in child care in recent decades (see chapter 3). Unlike routine housework, which goes down as paid work hours go up, mothers retain their child care time by cutting back on their own leisure, personal care, and sleep. This suggests that employed mothers are unwilling to delegate their child care duties. For example, highly educated women, who have more financial resources, average more child care time than do less educated women, and this also holds for highly educated men.[26] It seems that middle-class parents both feel compelled and are able to do more than do working-class families, as they subscribe more whole-heartedly to current parenting advice to "concertedly cultivate" their children.[27] While class differences in parenting style are discernible, compared with past times, parents are involved in more aspects of their children's lives, such as their education and friendships, and are less likely to let them play unsupervised.

Contemporary expectations about what constitutes proper parenting have thus intensified. Here I want to look in more detail at which child care activities parents perform and the distinctive temporality they involve. This will lead into a more general discussion about how to conceptualize the *slow time* of caring. It is precisely this special quality of caring time that sets limits to the automation of caring, the subject for the final section of this chapter.

Looking at time-use data from Australia, Denmark, France, and Italy, Lyn Craig and Killian Mullan set out to explore how routine housework is divided from nonroutine housework, and whether couples care for children in the presence of their spouse (copresence) or on their own (solo).[28] Child care was divided into two categories by activity type: (1) talk-based care, defined as face-to-face parent-child interaction that includes talking, listening, teaching, helping children learn, reading, telling stories, and playing games; and (2) routine, physical care and accompanying a child, defined as face-to-face parent-child interaction that includes feeding, bathing, dressing, putting children to sleep, carrying, hugging, soothing, transportation to school, visits, sports training, music and ballet lessons, school nights, meeting trains and buses, ensuring their safety, and handing them over to substitute caregivers.

National variations make a difference, so that, for example, average total parental child care time is highest in Australia and lowest in France because of the greater use and social acceptability of universal, state-funded early childhood education. Cultural attitudes about masculinity and fatherhood also matter, such that Danish men do slightly more routine care than fathers elsewhere, and Danish women do less child care solo. However, across all countries and household types, mothers do a significantly higher overall proportion of child care than do fathers.

Of particular interest here is the rare, quantitative data on the composition of activities. It turns out that fathers do only certain types of child care activity. "The widest gender gaps are in tasks that must be done regularly to a timetable, are less flexible, and, arguable, are less enjoyable than the talking, listening, reading, and play activities that constitute non-routine care."[29] Fathers' care time is spent in talk-based, educational, and recreational activities rather than routine physical and logistical tasks. Even more pronounced is the gender gap in the amount of care performed solo. Most of fathers' care time is done when the mother is also present. Laurent Lesnard, for example, reports that French fathers' time alone with their children is limited to a few minutes here and there, and largely consists of TV watching.[30] Conversely, much more of mothers' care time is done solo, making child care a more intensive activity for women, especially when children are young. It also means that fathers are not substituting for mothers' time, freeing them for other pursuits.

Caring time thus encompasses a wide range of activities and involves a complex set of emotions. The study above suggests that the quality of time may vary significantly with different kinds of child care. Talk-based care requires focused attention. It is less easy to combine with other activities than, say, routine tasks. In everyday life, however, the dividing line between domestic chores and parental responsibilities is nonexistent, as is the line separating family time, housework, and leisure. Leisure time on one's own has a different character than leisure in the presence of children precisely because the latter is a form of care. Negotiating overlapping and incompatible temporalities is therefore common as multiple activities are undertaken simultaneously. We are usually not aware of the coincidence of these different temporalities, so accustomed are we to integrating them.

Several authors have attempted to specify the distinctive temporal consciousness that characterizes caring. Such discussions emphasize that the dominant mode of temporality in modern industrial society, standard linear clock time, represents only one of many different tempos people inhabit. While abstract labor time is a medium of market exchange that can be bought and sold as a commodity, the more fluid open-ended times of the private sphere are harder to measure. "The expenditure of time in different economic spheres or social relationships may well be incommensurable. It is certainly not homogenous; nor can it be straightforwardly converted or "clocked" since there is no common external standard for conversion, other than clock time itself." Attempting to do this, argues Miriam Glucksmann, would be to confuse the measuring instrument with what it is intended to measure.[31]

Feminist theory draws attention to the embedded character of women's time. For example, Karen Davies introduces the concept of *process time* to describe the plural, relational, and context-linked nature of the time that caring for others involves. According to her, there is a clash between *nurturing rationality* rooted in the process time that good care requires and the *technical-administrative rationality* of workplace organizations.[32] Barbara Adam also stresses that, while both men and women are caught up in multiple times, time is not gender-neutral and many women's times are constituted in the shadow of the market economy. Caring tasks are often cyclical, fragmented, and woven into other processes rather than being completed as discrete tasks that can be accomplished.

While some early critics, such as Julia Kristeva, endorsed a dualistic opposition between female and male time, contemporary analyses maintain that "women's socially ascribed caring roles and, to a lesser extent, their physical role as reproducers, are linked to a range of temporal perceptions and logics very different from those that drive the labour market."[33] In this limited sense, it is meaningful to talk about *women's time*. As Valerie Bryson argues in her book *Gender and the Politics of Time*, we need to recognize temporal rhythms outside the commodified clock time of the capitalist economy in order to value time spent caring for others as an important economic and civic activity.

A gender perspective highlights how different kinds of tasks require different qualities of time and that speed, and the technologies that enable speed, may not enhance the relationship between the time spent and the activity. For instance, giving and receiving care involves slowness: "being there," as well as the emotional, affective dimensions of time. The politics of time goes beyond the redistribution of paid and unpaid work to include having time for ourselves as well as for others. More troubling for this literature on gender and time, it also asks us to rethink the relationship between care and work by foregrounding the pressures involved in straddling the multiple, contradictory temporalities of the public and private realm. The home is a domain where one ought to be able to regulate one's own time with a latitude rarely offered in the workplace. But is there something about the temporal modalities involved in family life that makes it especially hard to save time through automation?

Smart Houses and Caring Machines

A recurring theme of this chapter is the long-standing promise that domestic technology is the solution to the problem of housework. As with other forms of production, the future is projected in terms of automation eliminating the need for heavy and time-consuming labor. What hope is there that the smart house will finally deliver us from household drudgery? And will software agents or affective robots free us from some of the caring that slows us down?

Certainly the technological argument has intuitive appeal with regard to routine, repetitive tasks. Over the last decade, there have been a number of smart house and digital home installations and experiments in the United Kingdom, Europe, the United States, and Asia.

Yet prototypes of the intelligent house tend to ignore the whole range of functions that come under the umbrella of housework. Magazines like *Wired* and science fiction films like *The Matrix* present ubiquitous computing as the backbone infrastructure of the twenty-first-century lifestyle. Home informatics, much hyped as the Internet of Things, is mainly concerned with the centralized control of heating, lighting, security, information, entertainment, and energy consumption. My suspicion that designers and producers of the technological home have little interest in housework is confirmed by depictions of the MIT Media Lab's "CityHome," with its moving walls that enable the transformation of your bedroom into a gym.

With few exceptions, these visions of domestic life celebrate technology and its transformative power at the expense of home as a lived and living practice. The target consumer is implicitly the technically interested and entertainment-oriented male, someone tellingly in the designer's own image. The smart house they envision is more like Le Corbusier's "machine for living" than a home.

The range of smart appliances featured in the annual International Consumer Electronics Show reflects the attempt to find home applications for the functions that computers have excelled at in business and scientific settings—information processing and cataloging numerical processes. LG Electronics, for example, is developing a refrigerator that allows consumers to scan a grocery receipt with their smartphones so the refrigerator can track what is inside. So if you buy some chicken, for instance, the refrigerator will keep tabs on when you bought it and tell you when it is about to expire. If you have chicken, broccoli, and lemons in your refrigerator, it will offer recipes that include those three ingredients, even selecting recipes based on specific dietary needs and goals. Several manufacturers are introducing washers and dryers equipped with wi-fi that alert consumers on their television or smartphone when a load is done and gives them the option of fluffing the towels for another ten minutes. There is also a robotic vacuum cleaner with a built-in camera that can be operated remotely so that its owner can secretly watch what the nanny is doing.

Manufacturers' claims that this technological wizardry will make consumers' lives easier are hard to verify. While there is a market for smart technologies like the Roomba vacuum cleaner, for example, the variety and complexities of household labor impose limits on its mechanization. Even in the differently ordered world of paid work, robots perform only routine tasks in manufacturing, and much personal service work has proved impossible to automate. Domestic spaces are subject to a quite different set of considerations than those that govern the offices, factory floors, and workplaces within which information technologies have conventionally been deployed. While the dominant logic of capitalist workplaces is that of efficiency and profitability, a different logic governs domestic life—one that is primarily emotional and moral rather than quantitative.

It is surprising how socially conservative are the aspirations betrayed by the templates for the digital house of the future. The home of tomorrow is in fact an old idea and, like much science fiction, tells us more about our conceptions of the present than it does about likely futures. The most visionary futurists have us living in households that, in social rather than technological terms, resemble the idealized single-family household. The fact that "the world's 1.6 billion homes are as different from one another as the countries and cultures within which they were built" is at odds with the current, deceptively simple visions of the digital home.[34] The messiness of daily life is replaced with a vision of technological order. The space-age design effort is directed to a technological fix rather than engineering a less gendered allocation of housework and a better balance between working time and family time. The wired house may have much to offer, but democracy in the kitchen is not part of the package.

While automating routine housework is rarely the focus of technoscience, there is a vast research effort dedicated to computer software agents and robotics, no doubt because of their military applications. Nursebots, for example, are currently being developed for the commercial sector, for use in residential nursing homes for the elderly. However, as with other technologies, they are bound to be adapted and marketed for the home. So will these "caring" machines alleviate time pressure?

This again raises the issue of the multiple temporalities involved in caring. Nursebots, or mobile robotic assistants, can help to provide care for the elderly. They potentially offer assistance to escort people walking for exercise or to attend meals. These tasks are extremely time-consuming, because old people generally move at a very slow pace. Computers can also help monitor life signs and provide reminders regarding medication. Indeed, telemedicine is fast developing as a strategy for saving time and money in health services in the United States and Europe. However, many of the physical tasks that nursebots can perform simultaneously provide an opportunity for social interaction. When the "dead labor" embodied in machines is substituted for living labor, this opportunity is stripped away.

No longer. Computer scientists have been working on endowing robots with artificial intelligence for several decades. The field known as "affective computing" is extending this project to encompass emotional intelligence. The aim is for computers to be able to simulate emotions and feelings, to behave as if they have emotions. To this end, for example, a nursebot named "Flo" developed at the MIT Artificial Intelligence Labs was given basic facial features so that it could take an anthropomorphic form. Sociable robots in the form of Tamagotchi pets are commonplace in Japan. Indeed, Japan is in the forefront of care automation because it has an aging population and is a society with strong political resistance to immigration as a source of cheap domestic labor. (It should be noted that this project remains an aspiration of roboticists and their funders, rather than working technologies, as machines are yet to deliver the most basic forms of practical care.)

While innovations in robotics may well succeed in saving time, there are some who believe that it comes at too high a price. Sherry Turkle, who has written eloquently over many years about the creative potential of human-machine interaction, is worried.[35] Her book *Alone Together* sees humanity as nearing a "robotic moment" in which robots will be employed in caring roles, entertaining children or nursing the elderly, filling gaps in the social fabric left where the threads of community have frayed.

Human susceptibility to developing feelings of attachment to machines is a longstanding preoccupation of Turkle. Her studies of children and the elderly interacting with robots that imitate living companions reveal that, universally, a bond is formed. The Furby exerts a hold over anyone who nurtures it for a few weeks. More sophisticated models provoke deep emotional connections. Scientists developing the latest robots report feelings of pseudo-parental attachment. They hate leaving the machines "alone" in empty laboratories at night. "People attribute personality traits and gender to computers and even adjust their responses to avoid hurting the machines' 'feelings.'"[36] The extraordinary capacity people have for projecting human traits on to inanimate objects has been at the core of her writing. Now she is at pains to warn us that this intense relationship cannot be reciprocated. We are in danger of conflating caring as a behavior with caring as a feeling—machines can take care of us, but do not care about us.

While Japanese promotional material portrays the substitution of robotics for babysitting and housework as freeing up time to restore sociability, Turkle argues that the opposite is the case. She regards it as an irony that robots are being latched onto as the cure for a population increasingly isolated through the networked life. That is, our addictive immersion in digital connectivity means that time saved by robots will be spent plugged into the Internet and the mobile phone. Her view rests on the conviction that mobile phones, texting, and e-mail are creating a solipsistic universe; people are turning away from their family to focus attention on their screens. As I will demonstrate in the next chapter, this is a one-sided view of our relationship with technology in which we are being duped into increased dependency. Although this view has considerable currency, I argue that it fails to recognize the positive potential for cultivating a range of affective interactions between humans and machines.

Conclusion

This chapter has investigated why there is no simple connection between technology and time spent on housework. While domestic technologies have dramatically transformed our daily lives, they have not met the challenge of keeping home. Indeed, to take the example of the washing machine, it is notable that over the twentieth century the rate of innovation of this product has been very slow. Routine housework is still time-consuming and is still primarily performed by women; hence, solutions have been sought by buying in goods and services. Least amenable to automation is the temporality involved in sustaining the affective bonds of family life. Although, as we have seen, even these have been the target of robotics' designers.

It is interesting to note by way of postscript that analogous arguments have been made about earlier periods. Economists have long pondered why it is that some technologies diffuse at faster rates than others. Avner Offer, for example, is puzzled by the counterintuitive finding that household goods that use time (such as radio and television) diffused much more rapidly during the postwar years than appliances that save or optimize time (cookers, refrigerators and washing machines). Kitchen and laundry appliances lightened the drudgery of housework and contributed to its reduction, yet consumers gave higher priority to buying home entertainment technologies. His explanation for this "myopic choice" is in terms of people's preference for enhancing the immediate quality of their discretionary time over increasing its quantity. Moreover, "unlike the main housework appliances, 'time-users' delivered satisfactions directly to all members of the household, men and children as well as women."[37] While the gendering of artifacts does not explicitly figure in his model of consumption, the assumption here is that time-saving technologies were relatively devalued by men who had more purchasing power within households.

Offer's main thesis, however, is that the time saved by one set of appliances was consumed by the other. Entertainment appliances now claim the single largest category of leisure time use in both the United States and Britain. It is as if time saved by domestic appliances is spent watching the television screen. He views this as a myopic allocation of time, as the more television one watches the less satisfying it becomes. Indeed, he opines that television viewing has expanded to the point where it gives little more satisfaction than housework. As we shall see, the substitution argument (that new activities directly take time from previous ones) is problematic, as is the neat divide between time-saving and time-using goods, such as television as a form of child-minding. But Offer's overarching concern with the link between declining subjective well-being and time engrossed in compulsive television viewing chimes with Turkle's argument about our addiction to ICTs.

The extent to which our lives today are consumed by ICTs is without precedent, and will be the subject of the next chapter. I will argue that we need to move beyond such polarized views of the relationship between technologies and our experience of time. The theme running through this book is that there is no temporal logic inherent in technology so that even the same devices can have contradictory dynamics. We will see, for example, that mobile phones, widely blamed for accelerating the pace of life, play an important role in both maintaining sociability and providing a form of talk-based care. Contemporary communication may be highly mediated, but that does not necessarily mean that time spent communicating via these channels is of less quality or significance.

Chapter Six
Time to Talk

Intimacy through Technology

People's ecstasy concerning the triumphs of the telegraph and telephone often makes them overlook the fact that what really matters is the value of what one has to say, and that, compared with this, the speed or slowness of the means of communication is often a concern that could attain its present status only by usurpation.

GEORG SIMMEL, *The Philosophy of Money*

When Volkswagen announced that it would stop sending e-mails to employees after office hours, the *Independent* newspaper declared "victory for Volkswagen workers who just wanted to be left alone."[1] The article continued by emphasizing that the "tyranny of the out-of-hours e-mail from the boss has plagued workers the world over ever since the introduction of the BlackBerry. But now, after years of subjugation, one group of workers has struck a blow for freedom."

That our private time is being encroached upon by ubiquitous connectivity is widely regarded as driving the frantic pace of modern life. There are plentiful discussions about the inexorable extension of working time, reflected in terms such as work-to-family spillover, the colonization of time, and the blurring, merging and morphing of work-life boundaries. While the idea that the preoccupations of work can "spill over" into nonwork life is familiar, the ease with which digital devices teleport work into spaces and times once reserved for personal life represent a qualitative shift in the dynamic. The potential for work to invade every nanosecond is said by some to spell the end of pure, uninterrupted leisure time.

So what are the effects of the saturation of everyday life by instantaneous media?

The move to individualized, privatized communication on personal digital devices does mean that more and more of our social relationships are machine-mediated. Our personal time and private activities are punctuated—some say "perpetually disrupted"—by these communications. But interpretations vary between those who see this as increasing freedom and individual autonomy and those who fear a world of perpetual but less meaningful communication. Given how much of our lives is now spent on screens, is there still time to talk?

In chapter 4, I examined the impact of ICTs on the tempo of work at the workplace. In the first part of this chapter, I will focus on their role in redefining the boundaries between work and home and whether this is of more significance to feelings of harriedness than the pace of work itself. Taking issue with much of the literature in this area, I am going to argue against the view that ICTs extend and colonize all time outside the workplace.

In the second part of the chapter, I want to broaden the discussion beyond the all-too-familiar debates about the effect of ICTs on work-life balance. In my view, reframing such debates around the relationship between mediated communication and forms of close personal ties is a more productive way of thinking about these issues. This involves questioning the primacy typically accorded to direct, in-person interaction, which tends to obscure the role of material objects in the dynamics of affiliation. My own approach treats communication as embedded in various technologies that cannot be separated in terms of pure and mediated, with the former assumed to be more "real" than the latter. Digital devices can then be understood as fostering new patterns of social contact, providing a new tool for intimacy.

Multiple mediated forms of connection and distance, and the different negotiations that take place around access and availability do disrupt what we once thought of as boundaries between public and private, work and family, labor and leisure. However, rather than worrying about technology intruding on, and poaching, time from intimate moments that we used to think of as somehow "private," we would be better off reformulating the question in terms of the control that individuals can and do exercise over when and where they make themselves available.

I have been emphasizing throughout that time pressure is a multidimensional experience affected by a variety of factors. We have seen that economic imperatives, the rise of dual-earner families, and heightened expectations of parenting all accelerate our time frame. However, the temporal value of speed and efficiency and the desire to minimize the time in which a task is completed do not pervade every facet of our social and cultural life. While there is much emphasis on how technological devices accelerate activities and so crowd our time, I will argue that they can be allies in our quest for time control, preserving time as well as using it. The same features of ICTs that increase time pressure also offer flexibility in the timing and allocation of activities, and can facilitate temporal coordination. By intensifying connections with family and friends, ICTs can support new forms of personal ties that combine intimacy and spatial distance in newly configured times.

Families without Borders: Mobile Phones, Connectedness, and Work-Home Divisions

The division between the public world of work and the private domestic sphere has often been claimed as a distinctive feature of modern society. Arising in the middle of the nineteenth century but probably most fully achieved in the drift to suburbia in the middle of the twentieth, private life became the centre of new, secularized forms of self-fulfillment.[2] Postwar town planning reflected this, with cities and suburbs geographically segregated into zones for industrial and domestic use. Against this background, any threat to the inviolability of this personal realm is perceived as a risk to family balance, intimate relations, and personal identity.

So it is hardly surprising that the potential of digital technologies to dissolve the boundaries that once separated work and home life is the subject of much debate. Some sociologists even claim that "the distinction between public and private domains should be dispensed with since nothing much of contemporary social life remains on one side or the other of the divide."[3] Mobile phones are at the crux of these discussions, as they operate regardless of location, giving rise to a new pattern of continuous mediated interactions, variously known as *constant touch*, *perpetual contact*, *absent presence*, or *connected relationships*.[4] Many critics stress that, by allowing employers to contact their employees at all hours, mobile technologies encourage work problems to colonize the times and spaces once reserved for family life. Others, however, argue that by making place irrelevant, these devices afford novel opportunities for intense connectedness, deepening strong ties.[5] Rather than fragmenting relationships, time spent using communication tools might make relationships more durable.

What is at stake here is the effect of mobile technologies on the social organization of time. From the twentieth century onward, the regulation of working time has been a major method of social coordination, underpinning the capacity of all individuals to participate in joint leisure and recreation. This was reflected, for example, in twentieth century broadcasting, whereby radio and television programming schedules were timed to coincide with what was assumed to be a mass audience at home in the evening after work. Although the idea of a "normal" nine-to-five job still has a heavy hold on us, it no longer reflects the great diversity of hours or places in which people work. With the rise of flexible working hours and the predominance of dual earner families, reconciling the temporal regimes of paid work, leisure and family life has become increasingly complicated.

In this context, the traditional landline telephone looks anachronistic. Originally designed as a business tool, its use accorded with the notion that people demarcated and controlled their time outside of the workplace. Before the emergence of wireless telephony, there were of necessity separate phone lines (with separate numbers) for business and home. While using the fixed telephone at work was seen as a significant aid to employees in managing family issues while at work, most employers would limit access to home telephone numbers to "emergency" or crisis situations.

Digital communication devices have changed all that, providing people with the means to break from the traditional demarcation of work time from home time. All the evidence suggests that ICTs are indeed being used across social domains to accomplish work.[6] According to the Pew Internet survey, almost half (45 percent) of all working Americans report doing at least some work from home, although the segment that routinely works from home is smaller (18 percent). Checking work-related e-mail on the weekend has become routine for over half of all workers, and around one-third check their e-mail on vacation. However, the likelihood varies greatly by type of worker. Whereas well over two-thirds of those in the highest-earning occupations (earning more than $75,000) report working from home some of the time, less than one-third of those in jobs earning less than $30,000 a year do so. Those who own smartphones are much more likely to be required to read and respond to work-related e-mails when not at work.

Predictably, most of the existing studies about work spilling over into family time take a management or organizational perspective. The assumption is that the dramatic increase in workers' use of ICTs necessarily extends working time. For example, a Canadian study of managers, professionals, and technical workers actually defines cell phones, laptops, home computers, and BlackBerry devices as *work extending technology*—meaning the act of engaging in work-related activities outside of regular offices hours in locations other than the business office.[7] Some workers in this study, especially heavy BlackBerry users, did report a sense of having less family time as a result of working from home. Intriguingly, however, the main finding was that the very same features of ICTs that increase perceived control and facilitate communication are also the source of many oppressive features.

Noelle Chesley similarly found that frequent use of communication technologies by American professional and managerial career couples is linked to greater work-family permeability, lowering family satisfaction.[8] However, work-family spillover was found more typically among men, and family-work spillover more among women, with only women reporting that taking family-related calls at work was stressful. In this way, Chesley argues, these technologies may be reinforcing the gendered character of work-family boundaries, as family responsibilities appear more likely to influence women's outcomes. The potential for ICTs to simultaneously reinforce and rearticulate conventional gender scripts is a subject to which I will return.

The focus of these studies on high-earning employees bolsters the view that ICTs permit work extension, undermining domestic and family time. However, as I will demonstrate below, this is not necessarily indicative of more general societal trends in ICT usage.

A more fundamental problem, however, is that these studies take for granted the very boundary separation that they purport to be examining. The far-reaching changes wrought by digitalization call for a reexamination of this conceptual framework. *Boundary permeability* may not be the best way to frame the complex issues associated with a shifting work-family interface. As people further incorporate ICTs into the fabric of their everyday lives, it may be that the spatial, organizational, and even psychological distinctiveness of time at home and time at work will lose its salience. We therefore need to stop focusing on the impact of new communication technologies on the boundaries themselves. Instead, I argue, the task is to comprehend the practical ways in which these devices facilitate idiosyncratic patterns of availability, such that individuals no longer recognize much of the divide between these categories in the first place.

Much of the literature on spillover theory and work-life balance adopts a rather static model of the home as a fixed space in which family life is experienced and lived. By contrast, contemporary sociology of the family emphasizes relationships, connectedness, and practices. For example, David Morgan stresses that family life is always continuous with other areas of existence: "family practices are not necessarily practices which take place in time and space conventionally designated to do with 'family,' that is the home."[9] Rather, families are actively constructed through the day-to-day activities of their members, including in places of paid work. For Morgan, then, individuals are *doing* family, instead of passively residing within a pregiven structure. Family is designated less as a noun and more as a verb.[10]

The division between home and work, apparently so natural, is historically specific and built by social actors through repeated practices over time. Among these practices are those aimed at controlling the flow of information, communications, and demands across this boundary. The very concern about spillover and colonization signals the contested, changing nature of the public/private divide. These debates assume that mobile technologies inevitably produce workers, consumers and parents who are perpetually available. In the next section, I will start to show the limitations of such a view.

Patterns of Mobile Phone Use

To date, research on mobile phones has taken the form of either small-scale studies or surveys of attitudes toward the device.[11] My own research, which was conducted over several years starting in 2007, is based on a nationally representative survey of Australian employees. It was designed to investigate how individuals and households use the mobile to negotiate the different dimensions of everyday life.[12] The aim was not only to report on people's perceptions of their phone use but also to collect objective data on their actual behavior and practices. In order to capture this, I utilized two supplementary instruments: a unique analysis of the logs of calls held on mobile handsets and a purpose-designed time diary recording why, when, how often, and in what context people use their mobiles. Taken together, this information sheds light on the question of whether mobile phones help or hinder individual efforts to manage work and family time.

The common view, that the mobile phone promotes the colonization of personal time by job-related matters, is not supported by my findings. Indeed, major uses of the mobile, at any time of day, are not work related.

According to the nationwide survey, three-quarters of calls made on mobile devices and almost 90 percent of text messages are with workers' family and friends. Similarly, for almost all of the twenty-one thousand calls retrieved from handset logs, the predominant purposes are social. Thus, 40 percent of all voice communications involves contacting family, while staying in touch with friends accounts for a further fifth of all calls made and retrieved. A strong social emphasis is also evident in text messaging, with family (35 percent) and friends (25 percent) accounting for the overwhelming proportion of text messages. Only 21 percent of calls are work related, with men more likely to use their mobile phone for this purpose.

That the mobile phone is, in practice, predominantly a device for personal communication rather than work extension is underscored by the timing of calls. Both the phone log and the diary data show that work-related calls are mostly confined to standard business hours, rising sharply after 8 a.m. and declining around 5 p.m., with a small lunchtime dip. Importantly, work-related calls that could potentially extend work after hours (from 7 p.m. to midnight) constitute less than 3 percent of all the calls logged. Indeed, this low rate of work-related use out of standard hours suggests that something other than the mobile phone is extending work hours.[13]

Mobile phones, then, cannot be seen as primarily extending work and thereby fueling feelings of harriedness. On the contrary, by allowing some of the concerns of family and personal life to be handled during the working day, they might even be deployed to reduce time pressure. Over half of my survey respondents regard the mobile as helping them balance their family and working lives; very few report that it has a negative impact. When asked about the mobile's significance for family and household coordination, great importance was attached to information about the timing of the arrival at home and arranging to meet with other family members. Among parents, roughly two-thirds rated "arranging to deliver children" and "finding out where children are" as important, while planning meals was mentioned by a third. This softening of schedules, as well as time shifting activities to dead time, offsets—if not relieves—the extra time pressure created by potentially unlimited availability.

Much has been made of the mobile phone's role in *micro-coordination*,[14] the way it offers greater flexibility of schedules and control over timing, thereby saving time. Even making calls during dead time, which might sound like an escalation of the pace of events, might also relieve stress. The pattern reported above does indeed suggest that a major use of the mobile phone is for the temporal coordination of family affairs. Given the increasing complexity of managing the logistics of everyday life, the mobile phone may be the ideal tool for our contemporary times.

In a rare study, Emily Rose examines the specific ways in which employees engage in personal communication during the workday for these purposes. Instead of just noting the occurrence or not of such communication, she explores the intricate practices people devise in order to influence the extent and form of the connection between their work and personal lives.[15]

The mostly male engineers interviewed valued the ability to attend to matters that contributed to the organization of domestic life, as well as drawing on family and friends for personal and professional support. Equally though, they acknowledged that the unfettered flow of personal life into work was undesirable. Concerns included receiving unwanted communications and inappropriate messages and revealing too much of one's personal life to colleagues.

As a result, the employees deployed a range of strategies in a bid to exploit the benefits of personal contact during the workday while minimizing the downsides. They did this by managing, restricting or otherwise controlling incoming personal phone calls, texts and e-mail. This was achieved largely by granting prospective interlocutors access via specific modes. Some of these modes, such as e-mail and SMS, were favored because of their asynchronous properties, granting the receiver much greater freedom regarding when they respond. Moreover, in open-plan offices, the most discreet means of interacting typically is computer-based e-mail, whereby personal communication can be disguised as a work task.

Overall, the study found that employees actively manage the degree to which ICTs allow their personal lives to enter the workplace. While they take advantage of the possibility of engaging in personal mediated communication, at the same time, they engage in multifaceted strategies to restrict such permeability. This results in individually nuanced interfaces through which people from workers' personal lives have varying degrees of access to that worker. In this way, the engineers developed what might be thought of as a hierarchy of accessibility. While partners, close family members, and children are permitted to communicate by telephone (either landline or mobile), communication with friends is lower down the hierarchy, involving e-mail or texting. In other words, immediate access is denied to this group. Such research highlights the finely tuned ways in which workers utilize the materiality and functionality of technologies both to facilitate and impede contact with particular people at particular times.

In sum, fears that the constant availability afforded by mobile devices has accelerated the pace of life beyond people's perceived capacity to cope comfortably are exaggerated. Using mobiles does not straightforwardly lead to an increased sense of being rushed or pressed for time. More frequent use does not even lead to the experience of harried leisure. Indeed, it seems that the "always-on" character of digital technologies provides new opportunities for flexible coordination, counterbalancing any uncomfortable increase in time pressure.

Networked Families

At one level, then, portable devices enable us to organize more flexibly and efficiently and thus save time overall. Indeed, for Rich Ling, the mobile phone has completed the automobile revolution. "Where the automobile allows flexible transportation, up until the rise of mobile telephony there has been no similar improvement in the real-time ability to coordinate movements. When you were en route, you were incommunicado. The mobile telephone completes the circle."[16] So, for example, by redirecting travel that has already begun, the mobile reduces transportation time. Harking back to Lewis Mumford's famous line about the clock being "not merely a means of keeping track of hours, but of synchronizing actions," Ling argues that the mobile phone now competes with or supplements the wristwatch as the ultimate technology of micro-coordination.[17]

Throughout this book, however, I have been urging caution about making direct links between technological innovations and time saving. We have seen that there are many instances of artifacts, such as washing machines, that were supposed to save time. Evaluating such claims is always complex, as, to follow this example, people wash clothes more often. In other words, people commonly use technologies to enable them to achieve higher standards rather than to save time. There is a parallel with improvements in the speed of public transport. Many people have chosen to live further away from work as transport speed increased rather than to save time, so their commuting times stay the same.[18] The equivalent in the case of the mobile phone is that we may use it not so much to save time as to achieve more communications.

Furthermore, an increase in the volume of communication may result in communications of a different order. I have been arguing that we need to think about digital devices as sociomaterial practices that coevolve with lives as lived in interaction with technologies. If the social and the technical are mutually shaped, then technical innovations can foster novel patterns of social interaction, altering the quality of the times and spaces people occupy. In other words, mobile phones may be ushering in a range of new communication patterns, social relationships, and corresponding forms of life. It is thus fruitful to reformulate our analysis of work-home permeability in terms of how people are enacting these varying temporalities, and doing family, with digital technologies.

In this vein, Christian Licoppe suggests that ICTs provide a continuous pattern of mediated interactions that combine into "connected relationships," so that one can be present in time yet absent in the flesh. Rather than thinking about tasks and relationships as being located in one physical sphere or the other, he argues that new communication devices (such as the mobile phone) are not just added to other devices or substituted for rival uses. Instead, "it is the entire relational economy that is 'reworked' every time by the redistribution of the technological scene on which interpersonal sociability is played out."[19] Noting the frequency and short duration of mobile phone calls and mobile text messaging (in France), he argues that communication practices are being redirected toward connected interpersonal communication practices. Importantly, Licoppe stresses that this "connected" mode coexists with previous ways of managing "mediated" relationships, representing the emergence of a new repertoire for managing social relationships.

If connected relationships blur the experience of absence and presence, then by implication they are disrupting and rearticulating the public and private spheres. Having examined the claim that ICTs extend our working time, I now want to interrogate the related claim that they intrude on our intimate and family-based communication time. In other words, does the constant connection afforded by mobile modalities transform the quality of personal relationships? After all, families remain a crucial relational entity playing a fundamental part in the intimate life and connections between individuals. And building and maintaining relationships does take time together. So does the fact that we spend almost half of our waking hours using media and communications technologies leave us less time to talk face-to-face?[20]

Hardly a month goes by in which there isn't a new article or book on the subject of whether digital devices bring us together or push us apart. Just as in some early debates about the impact of television, many express the view that smartphones, tablets, and laptops are hollowing out our personal relationships and family time. In these scenarios, media use is portrayed as a replacement for, rather than a complement to, copresent family contact. For example, I have already mentioned Turkle's well-known argument that ICTs dilute the quality of time families spend together because individual family members are glued to a screen instead of socializing with each other.[21] Far from allowing us to communicate better, she views technology as isolating us in a cyber-reality that is a poor imitation of the human world. Addicted to multitasking on their BlackBerries, parents do not give children their full attention, and teenagers have become scared of the immediacy of talking on the telephone.

While some of the initial research did suggest that the introduction of the web increased loneliness, there has since been a plethora of studies showing that heavy users tend to have more, not less, social contact.[22] About half of British Internet users, for example, say that the Internet has increased the contact they have with friends and family.[23] And rather than replacing other forms of interaction with family and friends, the Internet complements other forms of contact. Most people communicate with family and friends at least once a week. Their means of keeping in touch are, in order of importance, face-to-face visits (84 percent), text messages (62 percent), telephone calls (60 percent mobile; 48 percent landline) and e-mails (33 percent). Around a quarter of adults use instant messaging on a weekly basis, and around one-tenth use mail to keep in touch with friends and family.[24]

Regardless of the communication modality, my research in Australia found that women are more likely than men to value keeping in touch with relatives, with almost 90 percent of women saying that the landline is either "important" or "very important," especially as a useful way of maintaining contact. The same gendered pattern holds for mobile phones and e-mails. This usage pattern also holds true for American women, who use landlines, mobile phones, e-mails and instant messaging more than men do.[25] They also contact children more often than men, especially using landlines and mobile phones.

Interestingly, this overall configuration is consistent with the earlier literature on the gendering of the landline telephone. Such studies show that a feminine culture of "kinkeeping," caring, and mutual support played a much more central part in women's use of the landline than in men's. As Lana Rakow reported from her American study of telephoning, women's talk is a form of caregiving, gendered work that women do to hold together the fabric of the community, building and maintaining relationships.[26] This emotional, or care, work is an aspect of intimacy underemphasized in mainstream writing on the family. Whether these gendered patterns are being incorporated into mobile phone usage is still an open question.

What is broadly accepted is that communication has become more central to intimate relationships. Anthony Giddens in particular argues that modern marriages and partnerships are based on a dialectic of mutual self-disclosure, a sharing of inner thoughts and feelings.[27]

Whereas traditional marriages were bound by economic and practical conventions, these egalitarian relationships are cemented by talk, relying on participants' willingness to continue because of their mutual pleasure therein. This profound cultural shift also characterizes parent-child relationships more than it did in the past. If intimacy is more and more characterized by rich communication, then the idea that communication technologies are somehow in opposition to "good" family relationships is contentious to say the least. Indeed, these devices may be providing an additional channel for performing intimacy.

In my view, then, the issue is to what extent physical, face-to-face relationships should be privileged in themselves. Certainly a particular form of copresence, choosing to spend time together to enjoy the pleasure of copresence, is a practice of intimacy in itself. Prioritizing time, offering privileged access to time, and seeking "quality time" are all ways of expressing affection, often within a spatially and temporally delineated private sphere. Does the fact that relationships with family and friends are increasingly digitally mediated mean that people place less value on being colocated?

Keeping in touch while physically apart is undoubtedly a marker of intimacy. The ability to be communicatively present with mobile devices may even enhance closeness at a distance. In order to gain some insight into this possible use, as part of my research I asked respondents, "If you and your partner are routinely apart for more than a day at a time, how important is the mobile phone in maintaining the quality of your relationship?" Approximately three-quarters of both men and women consider the mobile phone to be either very important or important in maintaining the quality of their relationship while geographically separated. Connecting with significant others, even via short calls or texts (phatic communication), can play a role in sustaining and deepening intimate relationships. Rather than conveying specific information, in many cases the call or text itself may be constitutive of the relationship.

Here and in my other writing on the mobile phone, I have argued that while people may erect a boundary in order to maintain a distance between work and home, they use the same property of mobiles—constant connection—to strengthen ties with kin and close friends at a distance. For example, when I did my study, nearly everyone was switching off their mobile phone in the cinema, two-thirds were switching them off at work meetings, and almost half turned them off in restaurants. However, only a minority turned their mobile off during leisure activities and during meal times at home. While one would now ask if people were turning their mobile phones to silent rather than switching them off, this pattern seems to indicate that people consider transgressing public spaces as more important than allowing calls at times reserved for family solidarity. Perhaps this is because the telephone is so closely associated with a deepening of connections with significant others that there is less need to control the flow over these temporal boundaries.

Indeed, people may positively welcome the softening of the separation between home and work afforded by these new devices, because, rather than fearing work intrusion, they are seeking deeper contact with family and friends. While the work-extension thesis emphasizes the dissolution of spatial and temporal boundaries, a focus on connected presence draws our attention to the social practices that constitute and maintain a private realm for affective relationships among family members and friends. This novel development reinforces the relational nature of family practices, deemphasizing domestic colocation and creating families without borders. Affective intimacy and the mobile phone are not in opposition to one another, as commentators of Turkle's ilk assume. Mobile communication devices should rather be regarded as another node in the flows of affect that create and bind intimacy.

Digital Youth

A major omission in my discussion so far is that I have not disaggregated abstract categories such as "family" and "friend" in order to examine how individuals in widely diverse personal relationships represented by these labels actually use ICTs. Just as in face-to-face talk, mediated communication has many more registers than those contributing to deepening intimacy.[28] Both face-to-face talk and mediated communication can be deployed in the exercise of control. Parents' (sometimes excessive) surveillance of their children is a case in point. It is both a practice of intimacy and a form of control. Structural inequalities between couples and between parents and children suggest that there will be a wide variation in how apparently similar calls and texts are interpreted. A single mobile call can have multiple functions, serving a range of purposes that are extremely difficult to categorize. There is considerable evidence, for example, that teenage children can experience mobile calls from their parents as a form of control and surveillance.

In relation to family time, studies vary in their findings. In Europe, for example, the work of Sonia Livingstone documents a trend toward privatized solo use of computer technologies among children consistent with living together and spending time part. She notes the advancing disappearance of family televisions and computers with the growth of individually used devices in the personal space of bedrooms.[29] Others disagree, suggesting that ICTs "have become the glue that binds some families together" through interaction between parents and children around the computer screen.

While much has been made of the decline in family time, in fact, the majority of families still spend considerable time together watching television, having family meals, and visiting friends and family. Almost all American adults who live with a partner or a child have dinner with members of their household at least a few times per week, and more than half do so on daily basis.[30] Moreover, parents, siblings, and other family members use media together in the context of home and family life. Over one-third of all parents say they play computer and video games, and most of these parents report playing video games with their children and that this brings their families closer together. Gaming and television watching (using digital video recording devices) are the most pervasive shared family activities.[31] And, according to the Pew "Networked Families" survey, nearly 90 percent of American parents say they have gone online with their children. More than half of married mothers and one-third of married fathers go online "often" with their children.

However, young people's experiences of intimate family time may well differ from that of the parents they are sharing it with. What mobile devices and social media have made possible to an unprecedented extent is for young people to express private lives *within* family time. Especially for teenagers or young adults, maintaining peer networks outside the home and family plays a vital role in the development of their sense of independence and autonomy. Mobile phones are integral to these practices and, arguably, they have changed the power dynamics of parental and peer group interaction for the adolescent. Indeed, these technologies give young people more control over their own time.

In particular, SMS, or text messages, provide adolescents with a form of interaction that they have adopted and shaped to their own purposes, turning it into "a living form of interaction." According to Norwegian researchers Rich Ling and Brigitte Yttri, texting's relatively low cost and the slightly illicit ability to silently maintain contact with friends (for example, during school or in the middle of the night under the bed covers) have been key to its popularity.[32] Crucial to its function is the physical nature of the phone—small and portable—and the fact that it is a personal communication device. That is, the mobile phone individualizes communication in that callers reach a specific person rather than reaching a random household member as with the landline telephone. Callers know that they are calling directly to an individual—not to a house phone where they may have to get through the filtering interactions of talking to a parent or another sibling. The mobile call is interpersonal and removes parental scrutiny.

In this way, the mobile phone provides young people with a means of being in touch with friends during times that were traditionally designated as times of family togetherness, such as during meals or on vacations. On one level this is not a new development, as other technologies like television have also permeated family space and reformed how families interact. Nonetheless, the authors conclude that the increase of mobile telephone usage is a distraction from the common focus of the ritual occasion, whether that is a common evening meal or the annual Christmas dinner: "the device steals attention from the shared family experience and thus limits its effect. The technology can undercut the potential for this type of solidarity and the hierarchical dimensions that this form of social structuring implies."[33]

As a counterpoint, I would emphasize that the speed and portability of digital communication may be enabling young people to maintain intimate friendships in new ways. Perhaps, in a distinctive manner, young people are now able to concurrently experience family time and time with friends. Time's elasticity means that it does not have to be apportioned in a unilinear mode. Rather than thinking about these interactions as exclusively family oriented or peer oriented, and setting up mediated communication in opposition to copresence, we need to complicate our understanding of the forms that intimate relationships can take in this digital era.

The notion that teenagers have an extended intimate sociality involving an always-on mode of communication emerges strongly from a major US study of the role of digital media and online communication in the lives of youth.[34] Social network sites, online games, video sharing, and gadgets such as iPods and mobile phones have become fixtures of youth culture and enable youth to connect with peers in innovative ways. While many parents and teachers interviewed for the study describe young people's hanging out with their friends online as "a waste of time," the authors argue that teenagers integrate new media ways of being together with the informal hanging out practice that have always characterized youth culture.

Notably, most of the direct personal communication that teenagers engage in through private messages, instant messaging, and mobile phone communication involves exchange with close friends and romantic partners, rather than the broader peer group with whom they have more passive access. Teenagers have a "full-time intimate community" with whom they communicate in an always-on temporal mode via mobiles and instant messaging. They use social network sites such as Facebook and Twitter to extend, enhance, and hang out with people they already know in their offline lives. Just as in the Norwegian study, the vast majority of teenagers use new media to reach out to their friends, casually connecting with them in private communication that is free from parental monitoring. While the expression of these peer social networks via mobiles varies among young people, cross-national comparisons reinforce a sense that in many parts of the world teenagers and children use mobile phones in broadly similar ways across cultures and countries and in broadly different ways as compared to adults.[35]

To summarize, ICTs do not impinge upon or steal private time in any straightforward way. Instead, they extend and reconfigure the time frames of those spaces, making possible new kinds of emotional proximity that are less anchored in shared time and geography. For young people, the so-called digital natives, omnipresent communication devices and social media stream seamlessly into the natural rhythms of daily life.

New Media Consumption

The salience for my argument of considering young people's use of new media is not solely because they represent the future. Doing so immediately calls into question the stereotype that everyone has busy lives. As Richard Harper remarks, "Teenagers don't bewail the fact that they have too many messages."[36] They communicate the most and yet delight in it.

Moreover, rather than always being used with an eye to saving time, mobile devices like television can fill time when there are few other options. Technologies provide an important escape route for youth who feel trapped in situations not of their making. Leslie Haddon and Jane Vincent's British study of children's use of mobile phones found that long car journeys were often cited as examples of phone use.[37] But children also cited times when they were with others—like parents—but wanted to be elsewhere. In such circumstances, using the mobile phone to play a game or listen to music are ways to mitigate a sense of boredom. It offers new sites for separation and autonomy, allowing the person to be present but also absent or withdrawn. Children send texts because there is nothing else of interest to do, as well as for positive reasons. For them, intensive media use helps time pass more quickly. The fact that adults, who receive far fewer communications, often complain of being overloaded reflects cultural narratives about ICTs speeding up life rather than simply being a function of the absolute number of mediated communications.

This narrative, in turn, is predicated on portraying past decades as less media infused and less rushed than our time today. But such comparisons are very hard to make. Let's take, for example, the time we used to spend watching television. Certainly, ownership of televisions has increased markedly, from 1.5 per average US household in 1970 to nearly 3 in 2010.[38] This is not necessarily a good indicator of how much time we actually spend watching television and how this relates to the time we now spend on new media.

A key issue is the extent to which new media displaces other ways of spending free time, and thus intensifies the pace of life. Yet there is little consensus about this. Even the figures on whether Internet use (for both communicating and informating) decreases television watching are disputed. According to most estimates, the amount of time given over to watching television seems remarkably large, on average accounting for about half of all leisure time. When this figure is added to estimates of Internet usage, there are not enough hours in the day for this to be plausible.

Measurement instruments, such as surveys by the World Internet Project and the Pew Research Center, time diaries, and automated monitoring on machines, all have different strengths but they do not match up.[39] For instance, according to time-diary data, there is no evidence that information technology leads to notable decreases in social life, media use, or other ways of spending both free time and nonleisure time, as claimed in earlier studies. As John Robinson remarks, this "'more-more' pattern is hard to explain and may point to the social desirability of doing more. In a society in which being busy is a 'badge of honor,' there may be societal pressure on survey respondents to report being more active than they actually are."[40]

Several additional factors are in play here. First, television watching is itself a very different experience than it was a generation ago. People watch television on various media, and they have much more control over what, when, and where they watch. Second, we know that in many homes television is used, rather like the radio, as background noise or wallpaper. It is on all day but does not require or receive exclusive attention. One can eat and watch television, talk and watch television, iron and watch television, and so on. Third, there is a strong trend toward multimodal connectivity and multiscreen viewing. A survey by Google reported that three-quarters of all TV viewers use another device at the same time in a typical day.[41] Moreover, whereas television, radio, and the landline telephone were essentially single-purpose devices, digital devices have a multiplicity of functions. None of the existing ways of measuring technology use listed above can hope to capture the rhythms of time into which ICTs are now so interwoven.

The question, then, is not so much whether new media accelerate the pace of life. Rather, we need to explore the degree to which the uneven diffusion of media usage throughout different aspects of daily life affects our sense of time. There are not enough hours in the day for all these practices, as they are not discrete, dedicated activities that one can measure in a linear way. If we move away from the binary opposition of mediated/nonmediated, then we can see that time consumption may involve media usage without being dominated by it. Just as debates on television viewing have become sophisticated, differentiating between diverse forms of television consumption, so too new media consumption should not be regarded on block as a unitary activity. The new devices have capacities that do not require full attention, and therefore time consumption is less linear. So we should not expect time-use data regarding all these changes to correspond with the segmentation of a standard day.

Work or Play?

Much of my discussion to this point has been framed in terms of work-life balance, and changing forms of labor is a recurring theme. However, rather than work intruding into leisure, social media are forcing us to rethink what actually counts as labor, and even what kind of experience labor is, in the first place. The fact that Internet users create most of the content that makes up the web brings into sharp relief our changing relationship to leisure time. After all, these new types of "digital," or "virtual," activities can feel like either work or play. Indeed, there are times when leisure is also labor.

From a political economy perspective, watching television advertisements and using Amazon, for example, can be viewed as forms of work in that they provide audiences and data for corporations. Tracking devices, such as cookies, record Internet users' browsing patterns and preferences, which can then be monetized through targeted advertisements. Online shopping is big business, with Britons leading the way spending on average over £1,000 a year, often on smartphones and tablets.[42]

Less visible is the free cultural and technical labor that drives innovation on the Internet: "simultaneously voluntarily given and unwaged, enjoyed and exploited, free labour on the Net includes the activity of building Web sites, modifying software packages, reading and participating in mailing lists, and building virtual spaces on MUDS and MOOs [multiuser text-based virtual environments]."[43] Indeed, there are millions of amateur programmers who are motivated by a desire for pleasurable cultural production, their only reward being the social status that comes with being at the frontiers of innovation.

In the case of audiences and advertising, labor is combined with pleasure in the sense that the nonadvertising content is viewed by choice. The delivery of entertainment content is thus a key component of the role of ICTs in work-life balance. Indeed, the life being balanced is not only the life of personal intimacy but of pleasurable recreation and relaxation. Spending time on Facebook, for example, can be read as a form of leisure as well as a form of communication. And while the traditional mass media of newspapers, radio, television, and cinema conveyed content to passive audiences, interactive media offer a much broader range of emotional and aesthetic expression, engagement and entertainment. By changing the relationship between the technology and the consumer in this way, ICTs arguably have transformed the quality of our leisure time.

This is not the place to delve into the extensive academic debate about the effects of new media on the nature of consumer culture. It is, however, worth pausing to consider just how much personal digital devices and systems facilitate and foster user involvement.[44] Throughout the book I have treated the Internet as a sociomaterial practice that is open to manifold usages, although, as I have emphasized, its network architecture is shaped by powerful commercial interests to steer users down particular pathways. Here I have been focusing on the ways in which ICTs reconfigure the temporal and spatial basis of intimate relationships. But developments such as Web 2.0, open-source software, and Wikipedia show how the unique materialities of ICTs are also generating unprecedented and unpredictable cultural and informational practices. Indeed, the web is itself animated by users who are both producers and consumers. In this sense, the Internet is quintessentially a "generative" technology in a way that no previous technology has been.

So argues Jonathan Zittrain in *The Future of the Internet*. He sets out to compare the generative qualities of previous tools, such as hammers, Lego bricks, knives, and dumbbells with the Internet and traditional PC architecture. While the former are versatile and can be combined in a variety of ways, truly generative systems contain the capacity "to produce unanticipated change through unfiltered contributions from broad and varied audiences."[45] Compared to traditional technologies, PCs and network technologies produce both technical innovation and participatory output, albeit some of it unwelcome (such as malware).[46]

Crucially for the time-pressure paradox, generativity gives rise to new channels for social networking and creative expression, new forms of work and a host of new preoccupations and activities. Such changes are happening at a startling rate. For example, a multibillion-dollar apps industry has arisen virtually overnight. The iPhone and iPad already have about seven hundred thousand apps, from Instagram to Angry Birds, and the number of software engineers, including freelance app writers, has mushroomed to over a million.[47] Rather than being confined to massive R&D labs, the process of developing and making hardware and software is much more distributed.

More broadly, the Internet makes it possible to reshape "both the 'who' and the 'how' of cultural production. . . . It adds to the centralized, market-oriented production system a new framework of radically decentralized individual and cooperative nonmarket production."[48] Individuals and groups can actively participate in the process of cultural creation. This is not to suggest that the Internet guarantees an "interactive society," as some would have it, nor to imply that broadcast media are incapable of interactivity.[49] But there is little doubt that digital media open up new opportunities for the democratization of cultural production, providing important sites of self-expression in the twenty-first century. The fact that human-machine interactions characterize so much of our leisure time needs to be understood as adding a dimension to temporal experience. It certainly confounds the dynamics of when we are at work or play. The speed or instantaneity of contemporary consumer cultural is a further issue, which I will take up in my final chapter.

Conclusion

From the telegraph to the landline telephone, communication technologies have long shaped social interaction. However, since the advent of the mobile phone and the Internet, interpersonal relationships are routinely conducted via machines in times and places once regarded as the public domain. The customary split between the public and private is now overlaid with and complicated by another layer of electronic network connections. Sharing our most intimate lives in multiple sites beyond the physical space of the house challenges our very conception of the home. Activities that once defined the separate times devoted to family, work, and leisure are being combined in new constellations. It is my argument that digital devices open up fresh possibilities for what we can do with our time that goes beyond the fixed distinctions between work and family.

Reformulating the issue in this way renders the common portrayal of multimodal machines colonizing our time and starving us of time to talk, as distorted. Indeed, constant connectivity has largely been welcomed, for enhancing relationships with family and friends as well as increasing contact with wider social networks. Such mediated communications do not replace embodied forms of mutual presence. Rather, these sound and screen modes of sociality coexist alongside face-to-face interaction, building new levels of connected engagement. What is striking, even among young people, is the extent to which social media like Facebook involve members who meet in the offline world. This both affirms the distinctive quality of time spent in shared presence and suggests that the additional channels of multimedia enrich embodied encounters when they occur.

Inevitably, the constant flow of communication requires negotiation over the allocation of time and attention in multiple temporal zones, causing communication congestion and conflicts. Making sense of these devices and developing norms for their habitual usage is, after all, an ongoing process involving a continuous feedback loop. Nowhere is this more critical than in relation to the ease and speed with which ICTs allow some kinds of paid work to leak into what has hitherto been regarded as our free time. Fears about 24/7 electronic connectedness have to be understood in the context of the harsh economic climate and its attendant insecurity.

Nevertheless, the same technologies that promote the extension of working time can also increase autonomy and control over when and where work tasks are accomplished. In chapter 3 I argued that one of the main causes of the time squeeze is temporal disorganization, the increased difficulty of scheduling and synchronizing social practices with others in a deroutinized society. Mobile phones in particular offer flexibility in the timing and arrangement of activities, thereby facilitating temporal coordination. In this way, ICTs can be a potent resource for augmenting work-life integration and deepening intimacy.

I will end by recalling a recent visit to my ninety-five-year-old mother in a nursing home in Melbourne. In the garden I witnessed a frail old woman, silent and slumped in a wheelchair next to her daughter, who sat with one arm around her mother while using a state-of-the-art smartphone with her other hand. My initial reaction to seeing her tapping away was rather negative, as clearly her attention was not on her aged mother but elsewhere. As I observed for longer, however, it was apparent that her mother had little consciousness of her surroundings but was probably comforted by her daughter's arm around her. The timescape of nursing homes is extremely slow for both patients and visitors, and much of the care administered involves filling time and being present. It is not amenable to acceleration by technology. Given this context, perhaps the daughter was smart to be making the most of technology to traverse different time zones.

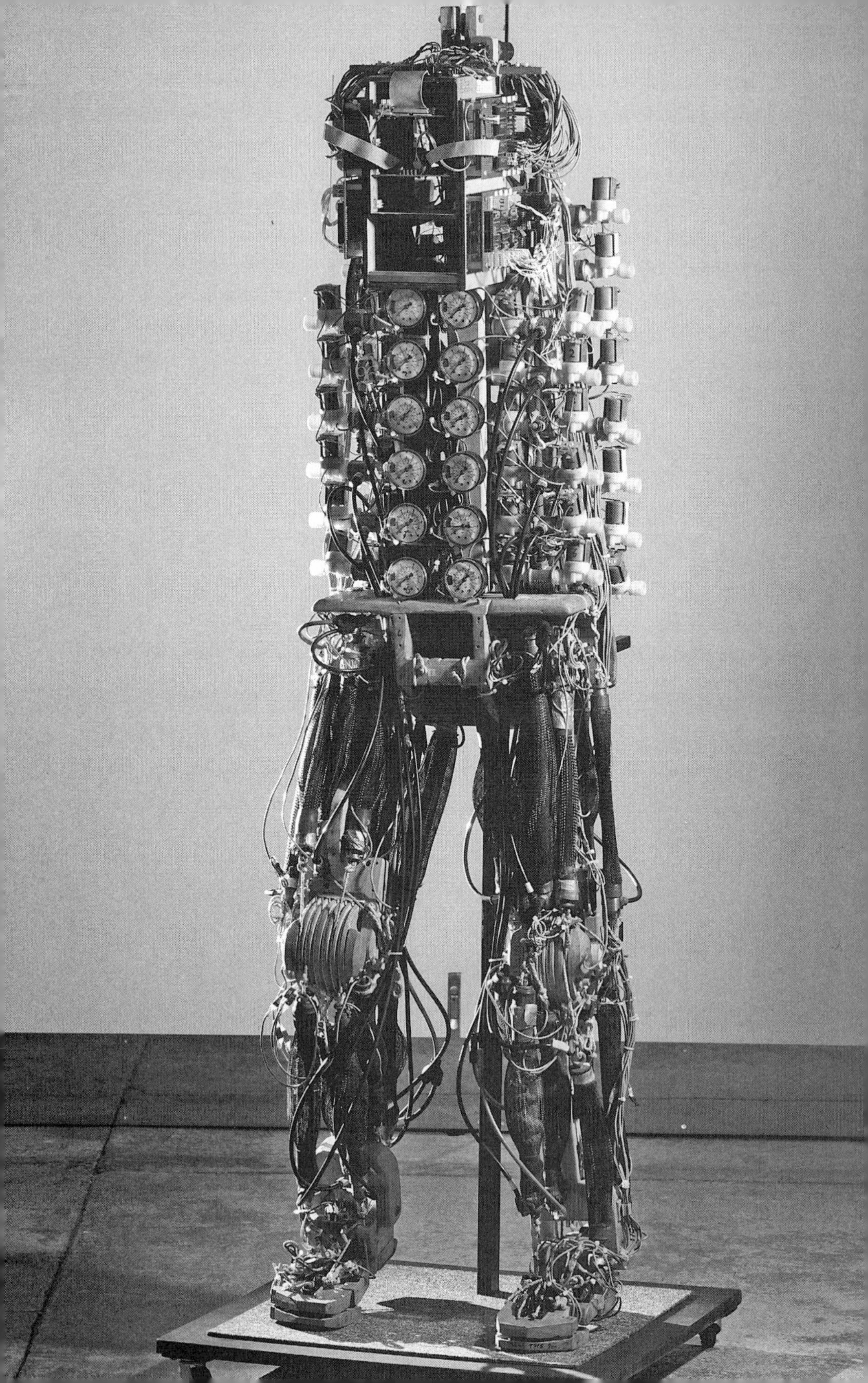

Chapter Seven

Finding Time in a Digital Age

We need to de-alienate time: reconnect clock time to its sources and recognise its created machine character. As such, concern with the multiple time dimensions of our lives is no mere theoretical, academic exercise; rather, it is a strategy for living.

BARBARA ADAM, *Timewatch: The Social Analysis of Time*

Almost a hundred years ago, economist John Maynard Keynes imagined that by the beginning of the twenty-first century, we in the West would only have to work three hours a day to satisfy all our needs. He anticipated that the constant growth of productivity resulting from technical progress would eventually solve the economic problem of supplying humanity's material needs at a fraction of the existing work effort. The abundance of time thus released, Keynes hoped, would lead to a moment when the spontaneous, joyful attitude to life then confined to artists and free spirits was diffused throughout society as a whole.[1]

Yet the reverse seems to be true. Machines have not liberated us from work in the way Keynes predicted. The unparalleled velocity of computerization, telecommunications, and transport, which was expected to free up human time, has paradoxically been accompanied by a growing sense of time pressure. Rather than inhabiting a world in which time is abundant, everyday life seems more rushed. Although the details of time scarcity vary substantially across socioeconomic groups, as a culture we have a shared experience of temporal impoverishment. This is the conundrum that I have been exploring, that we live in an *acceleration society* in which technological acceleration produces not more leisure and downtime, but in fact an ever-faster pace of life.

Here, I am interested in how we can make more of time. It is not literally possible to make time in the sense of adding another hour to the day. Rather, the key to understanding the fraught and complex relationship between technology and time is the concept of temporal sovereignty, the ability to choose how you allocate your time. I have argued that having discretionary control or autonomy over your time lies at the core of some positive notions of freedom. It is a significant measure of life satisfaction and well-being. The proliferation of hyperefficient ICTs that should help us take control of time, seem instead to control us. As in the classic *Frankenstein* myth, it is often lamented that we have lost control over the machines to which we gave birth. We wonder whether a faster life, replete with gizmos, is necessarily a better life when the trade-off between time and money seems to turn every hour into the rush hour.

Certainly it is hard to disentangle the profound social changes that have occurred in our lifetime from the closely interconnected transformations in technological systems. However, it is only by examining the acceleration society thesis in some forensic detail that we can identify which aspects of life are accelerating, which are slowing down, and for whom. My social shaping framework has undermined the notion that the acceleration of work, parenting, and leisure is directly driven by technology. In tandem with technological change, I have shown that major shifts in the nature of work, the composition of families, ideas about parenting, and patterns of consumption have all contributed to our sense that the world is moving faster than hitherto.

The experience of time pressure, or harriedness, then, is not simply a function of machine speed. It is therefore not amenable to resolution by means of a digital detox diet: "periodically shut down the electronic prostheses dictating our worlds and lives. . . . Shut off the cell phone; ignore e-mail; disable the answering machine and caller ID."[2] In any event, I do not subscribe to the nostalgia for a more natural, less technologically suffused past that some advocates for slow time champion. Rather, returning to Donna Haraway, I want to embrace the emancipatory potential of technoscience to create new meanings and new worlds while at the same time being its chief critic.

The design of our devices and the material infrastructure we inhabit reflects, as much as shapes, the society in which we live. It embodies the dominant engineering approach to time saving and time ordering, a particular conception of efficiency. Likewise, the promises and cultural imaginaries of a technological future populated with robots and autonomous software agents project limited visions of a good society. Opening up the processes of technical innovation to encompass a broader range of societal realities and concerns is the only way to generate new devices for new times.

Contesting the contemporary time culture, then, cannot be thought outside of and separately from technological developments and vice versa. We make sense of and operate in the world together with the machines of our making. And while instrumental time is built into these material things, there is a complex dialectic whereby increased technical speed can simultaneously provoke new slowed timescapes.

In this final chapter, I want to remind you, the reader, of the diverse ways in which technology reconfigures time, and gesture toward some possible directions for making more of time. There are two broad aspects to this time quest. Both involve shifting the emphasis away from how digital devices colonize our time to a more political orientation based on how time is allocated and how it is valued. The first issue is the familiar one of reducing working hours in favor of leisure time. This terrain has become harder to navigate as ICTs dissolve the boundaries between home and work. The second aspect is somewhat speculative, and involves examining whether we can alter the texture and tempo of life and what role technologies might play in this. I will consider these in turn.

Reformulating Working Time

A young colleague of mine mentioned that he was using a time-management app in order to use his time more efficiently. Apparently the smartphone application enables you to track exactly what you do with every minute of the day. As far as I know, he is not signed up to the quantified self movement, whose members use personal tracking technologies to monitor their every move.[3] It hardly needs saying that this form of self-auditing is a highly individualized response to collective problems, but for him, the latest devices are a powerful resource that enables him to take control of time. Such apps are based on the well-worn time-management belief in the unlimited virtues of acceleration, that we should do everything faster. In other words, it reduces all time to a standard metric. Wasting time is bad and we should maximize our productivity.

How we use our time is fundamentally affected by the temporal parameters of work. Yet there is nothing natural or inevitable about the way we work. As we saw in chapter 2, the idea that labor is measured and regulated by linear clock time is a relatively recent feature of industrial societies. The quest for maximum speed and efficiency, the disciplined and frugal use of time, only became hegemonic in a market economy where time is money. Fewer workers clock in and out of their jobs these days, but the logic of industrial time still ticks away, shaping how we understand our lives.

At one level, the most straightforward way to alleviate time pressure would be to reduce working hours. We only need to remind ourselves of the long, grinding working hours of previous eras to see how far we have come. Hours of work per person fell rapidly from 1870 to 1930, and Keynes assumed that this fall would continue. The establishment of the standard eight-hour day and the five-day week in the decades after the Second World War was a landmark achievement of twentieth-century social democracy.

However, this trend to shorter working hours has stalled and, for some, it has gone into reverse. In *How Much Is Enough?*, Robert and Edward Skidelsky take up Keynes's challenge and puzzle over why: "we in the rich world are four or five times better off on average that we were in 1930, but our average hours of work have fallen by only a fifth since then."[4]

Their explanation for the continuation of long working hours is twofold: a capitalist economy gives employers the power to dictate hours and terms of work, and such an economy inflames our insatiable desire for consumption goods. At its root, however, our addiction to work and hyperconsumption is due to the disappearance from public discussion of any idea of the good life in which leisure would be valued for its own sake. The sociologist Juliet Schor similarly puts a new allocation of time at the centre of *Plenitude: The New Economics of True Wealth*.[5] Like the authors above, she rails against the long-hours culture, limitless growth, and overconsumption. According to Schor, we need to revalue well-being, as millions of Americans have lost control over the basic rhythm of their daily lives: "they work too much, eat too quickly, socialize too little, drive and sit in traffic for too many hours, don't get enough sleep, and feel harried too much of the time."

Such arguments for reducing the incentive to work are based on the *average* number of hours people work. The authors are well aware of the growing disparity between the work rich/time poor and those with few or no hours of work. Indeed, they are motivated by a concern with a fairer distribution of working hours, ending the divide between those who are compelled to work longer than they want, and those who cannot get enough work.[6] Nonetheless, their general point stands: the culture of today's opulent societies is more harried, not more leisurely.

Both books reflect a renewed interest in the politics of working time. I was particularly struck by this, as my own introduction to the issue was via earlier leftist arguments for a leisure society based on affluence.[7] By contrast, current critiques of endless growth are inflected by the economic downturn and the need for environmental sustainability. And instead of the socialist glorification of work as the key to collegiality, solidarity and democracy, these authors question our willingness to dedicate ourselves to a life of relentless work. In doing so, they recall ideas of leisure advocated not only by Keynes and his contemporary, the philosopher Bertrand Russell in his essay *In Praise of Idleness* (1932), but earlier by Marx's son-in-law Paul Lafargue in *The Right to be Lazy* (1883).[8] Unfortunately, in common with their predecessors, this vision of a progressive politics of time seems only able to invoke a peculiarly bourgeois and suburban ideal of home and family life.

Missing from these attempts to reformulate working hours is any recognition of the gender dynamics embedded in how we think about time and work. A fair distribution of work would need to take into account not only the different patterns of employment currently found between men and women, but also the inequitable distribution of unpaid work *within* households. While the authors above correctly criticize the framing of time in terms of productivity growth for its own sake, they do not question the differential worth accorded to the public use of time spent in paid work, compared to private domestic labor. Kathi Weeks reminds us that the Marxist productivist paradigm undervalued and rendered invisible the labor time of unwaged housework, caring, and emotional work. A feminist time movement "should attend to the whole of the working day by, for example, insisting that estimates of the socially necessary domestic labor time of individuals be included in both calculations of working time and proposals for its reduction."[9]

It is precisely this combination of paid and unpaid work that makes time poverty so widespread among working women. As I have already elaborated, a major cause of time pressure is the growth of dual-earner households, which supply more working hours to the labor market than ever before. Especially in the context of the intensified expectations of parenting, mothers in full-time employment are particularly busy juggling the conflicting demands of work, family, and leisure.

This is not only a matter of the length or duration of time. Caring and attentiveness cannot be reduced to linear time as though they involve a sequence of tasks that in principle could be delegated to machines. What is colloquially referred to as "quality time," or being temporally present with children, requires a cadence that is not subject to acceleration. Just as one cannot ask an orchestra to play twice as fast as the score requires, the character or intensity of giving and receiving time contributes to the experience. For example, we have seen that women's leisure time is "less leisurely" than men's, as women are more likely to combine leisure with looking after children. What presents itself in public as a discussion about a reduction in working hours conceals the ways in which timescapes are still differentiated along sex-specific lines.

However, even this critique of the gendered nature of work does not go far enough. Attempts to revalue domestic labor still operate as if the distinction between the public and private domains can be clearly delineated. Technology barely figures in any of these discussions about the politics of time, except as an external factor that eats into leisure. But the pervasiveness of ICTs into every aspect of our lives poses the profound question of whether the dichotomy between work and personal time still holds in a digital age. This issue is at the core of the concerns that I expand on below.

Work-Life Articulation

We cannot have a serious discussion about working time today without interrogating the way that ICTs confound the distinction between "my time" and "work time." Just as the standard five-day work week has been heavily eroded, so too have the time and place of work. Whereas the old industrial clock regulated our lives in discrete blocks of time and space, with the separate spheres of public and private life, the constant connectivity and global reach of mobile, digital technologies erase time zones and specific work places. The traditional time/space of the week and weekend and their characteristic social relations are now porous as people increasingly work, play, consume, and interact anywhere, anytime.

Throughout the book I have argued that ICTs create a multiplicity of temporalities and modes of everyday living, as people craft new understandings of themselves and their relation to others. Time cannot be thought of as an abstraction, divorced from a socially situated materiality and embodiment. We make and measure time with and through instruments, tools, and techniques. But it does not follow that accelerating technologies inevitably hasten the pace of all social domains. The way artifacts evolve in relation to time practices crucially depends on how they become embedded in our institutions and the vicissitudes of ordinary life.

The very same machines that can make us feel harried also free up time, allowing for much greater autonomy, flexibility, and versatility in how we organize human affairs. Recall our discussion of the smartphone, the quintessential time-space compression mechanism. Not only does it both save and consume time, it also transforms linear, sequential timescapes, and all this simultaneously. I have argued that the experience of harriedness takes a variety of forms, depending on which aspect of temporality is being squeezed. Problems of temporal disorganization, the difficulty of coordinating shared social practices with others, loom large with the growth of dual-earner households. So it is not surprising that people have actively embraced and appropriated cell phones in order to microcoordinate and synchronize their multifarious activities. By softening schedules and making time more fluid, these devices offer unique techniques for alleviating this aspect of time pressure.

Indeed, I have debunked the notion that we have all become cyberserfs, technologically tethered workers with no control over our own lives. In chapter 4 I considered the complex entanglement of contemporary work practices, working time, and the materiality of technical artifacts. I showed that neither the smartphone nor even the sheer volume of e-mail traffic drives the speed of work. People overloaded with work reach for their electronic gadgets in an attempt to relieve the pressure that the devices magnify but do not in themselves cause.

This is not to deny the distinctive materiality of digital technologies and its powerful effect on organizational practices. The speed of e-mail, for example, promotes constant connectivity and instant response. This technical affordance has agency in that it contributes to the naturalization or taken-for-grantedness of this practice.

It is precisely because of this that there is a tendency among advocates of shorter working hours to presume that quality personal time entails escape from the encroachment of electronic paraphernalia. But in my view, digitalization provokes a radical rethinking of the standard terms of the work-life balance debate, which pitches work against life and public against private. ICTs make possible new combinations of previously distinct temporal zones, new forms of mediated intimacy, and new ways of doing family. Mediated relationships have not supplanted embodied mutual presence, but rather exist alongside them. The phenomenological experience of being in the presence of others at a distance can invigorate and intensify communication, rather than detract from it. Contrary to much of the hype, we may well have more time to talk.

Digital technologies, then, must be understood as more than simply tools for exchanging data and coordinating human interaction. They do not just make existing forms of social action more efficient. As material objects or sociotechnical assemblages, they reconfigure the temporal and spatial dynamics of how people think and act. People may welcome the permeability of these boundaries for the flexibility and control it offers rather than primarily fearing work intrusion into leisure time.

The important questions for a politics of time, in our age, might have less to do with a sense of harriedness or the mixture of work and home life than with a hierarchal time culture in which status and pay measure the value of a person's time. Being busy is valorized, while having too much time on one's hands signifies failure. Temporal disparities are closely mapped to social inequalities, as exemplified by the demonization of the unemployed. The democratization of time would lead to a very different social order, one in which time priorities and restraints are equitably shouldered. I would not be the first to suggest that "the concept that everyone's time is equally valuable is truly revolutionary."[10]

Recognizing this seems to me to be a first step toward a fairer distribution of work.

Hectic Leisure in a Self-Service Economy

Up to this point, I have focused on the length of working hours. A related but even knottier issue is the tempo of activities themselves. I have already alluded to the intensification of work and the demands of multitasking required of today's networked workers. But what about the acceleration of other aspects of our lives, such as the instantaneity of consumption and the intensity of leisure? If we had more time, how would we make the most of it?

Leisure time is suffused with ICTs, and I have talked about the trend toward multimodal connectivity and multiscreen viewing. The vast expansion of the market for consumer electronics, such as computers, cell phones, televisions, tablets, and MP3 players, is paired with an unprecedented speed of built-in obsolescence and product abandonment. We expect instant delivery of products of all kinds, and then discard them with equal speed, typically without a thought about the conditions of their production or the waste they leave behind. Indeed, one reason that time seems scarce is because it is impossible to consume the vast array of products and services on offer. Whether or not we view human desires as inherently insatiable, contemporary consumer culture is certainly characterized by excess. Symptomatic of this is the endless upgrading of smartphones and a World Wide Web that can be surfed indefinitely.

One rarely noted way in which technologies consume time is that their rapid cycle of renewal requires an ongoing investment in skill acquisition. Familiarizing oneself with and learning how to operate digital devices requires substantial inputs of unpaid user time. Internet use, for example, requires an infrastructure and its maintenance. As the apocryphal story goes, if General Motors automobiles seized up as often as Bill Gates's software, we would certainly not rate them. Our lower expectations of software have been well managed by marketing, so that we often blame ourselves for its deficiency. Even shopping online, which promises saving time, can be a tiring activity squeezed into time once reserved for rest and leisure. Automated telephone answering systems are the subject of many jokes that express our frustration at being at the mercy of a technology that is saving someone else's time and money at our expense. Consumption in an instant society can, at times, involve a temporality of remarkable slowness.

The ascribed nexus between high-tech and efficiency is often belied even in conventional, in-the-flesh shopping. Not long ago I visited the Apple store in central London to repair my iPod. As it is over five years old, I was firmly informed that no spare parts existed for such aged machines, that it was redundant and I would have to buy a new one. OK, I said, bring me the new model in any color and I will buy it—where is the cashier? I look around but there was nowhere in the shop to purchase a product. Apple policy is a "personalized" service, so you have to wait for someone to serve you individually. In fact, it took ages. When I inquired as to why there were no such counters, I was told that the company had dispensed with them in order to avoid the appearance of long lines! In their attempt to abolish the cardinal sin of waiting, the company had inadvertently designed a slow service as the condition for purchasing the latest, fastest product. Slowness as a common condition of contemporary urban life is rarely diagnosed.

The rise of the self-service economy, whereby more and more of the work of consumption is transferred onto consumers, was first theorized in the 1970s.[11] ATMs, self-service gas stations, and vending machines were spreading and promoted as being time saving for consumers. More recently, self-service tills at supermarkets and preprepared meals are on the increase. No one could have anticipated the huge growth of online shopping, let alone 3D automated manufacturing or self-driving cars. This trend is being taken to new heights by the world of Big Data that promises a future of frictionless, continuous shopping. As we are tracked and fed information on our wants and needs from online tools, services and apps, we are assured that our consumption will only get more streamlined and efficient. We will no longer need to search for anything since we are perpetually monitored, with the relevant information sent to us on the basis of perceived need.

Amazon, Google, and Wal-Mart are all moving toward same-day shipping, as people become accustomed to instant gratification, where you just type something on your phone and the next thing you know, you have what you need.[12] This will soon include Sunday deliveries, boasts

Amazon's Jeff Bezos, with the aid of drones. Dinner kits are also taking off. E-commerce businesses, such as Plated, will buy, measure, cut, chill, box, and ship every ingredient for a meal to your door. All the customer has to do is to order online. As one of their spokesmen put it: "Food is one of the last pieces of daily life that is still analog. We want to bring it into the digital age."

Ironically, some of these futuristic developments hark back to the past. During the nineteenth century, household deliveries and mail-order catalogs were common. As I outlined in chapter 5, the industrial revolution in the home and the advent of the car, paradoxically, increased shopping time. For all the e-commerce today, shopping and domestic travel times show a rising trend. Even sophisticated domestic appliances, such as the washing machine, have been more successful in time shifting and raising standards than in reducing housework. Rather than compressing labor time, such machines radically alter expectations of comfort, cleanliness, and convenience. Routine household work is consequently still time-consuming, and that is why the cash-rich, time-poor employ paid domestic workers and fast food outlets flourish on busy main streets.

In the end, the relationship between technological change and temporality is always dialectical: the simultaneous production of fast time spaces with those of remarkable slowness. Speed and slow down have always coexisted in modernity, although the meanings and values attached to them have shifted.

Notwithstanding this recursive interplay, the predominant emphasis in social and cultural theory is on the acceleration of everything. As outlined in chapter 1, this new temporality is variously described as one of immediacy, instantaneity, simultaneity, timelessness, chronoscopic, or network time. Ben Agger has gone so far as to label it iTime, a manic, compulsive, deeply compressed time "weighing heavily on the person who always has too much to do, not enough time to do it."[13]

For John Tomlinson, too, this telemediated culture is a novel historical phenomenon, a unique way of being.[14] "Immediacy"—the combination of fast capitalism and the saturation of the everyday by media technologies—changes the nature of consumer culture entirely. It is characterized by instantaneity, proximity, and expectations of immediacy that are linked to "assumptions of instant delivery and effortlessly achievable abundance." Accordingly, "delivery" itself rather than satisfaction becomes the *telos* of consumption; "something new is always on the way and so it is not necessary—nor does it do—to invest too much into the thing of the moment." While Tomlinson highlights the usual critiques of accelerating consumption, such as debt, waste, and maintenance of the capitalist status quo, his real concern is that the culture of immediacy is incapable of generating "new imaginations of the good life."

Slow Living in Modern Times

It cannot go unremarked that, alongside the reverence accorded to a hyperkinetic, digitally fueled pace, the idea of a slower life is gaining appeal. Recent decades have spawned several slow-living initiatives, such as the Slow Food movement, Slow Cities (CittaSlow), the Society for the Deceleration of Time, the Simple Living Network, and various forms of meditation and mindfulness. Indeed, there has been considerable growth of the "mind business" within large corporations.[15] There is even a manifesto for Slow Science, calling for scientists to proceed with less haste, taking time to think, to read, and to fail.[16] Common practices associated with slow living include cooking and sharing a meal instead of buying fast food, growing fruit and vegetables locally, and cycling or walking instead of driving. These initiatives deliberately posit slowness or deceleration as a subversion of the cultural orthodoxy of speed.

Clearly, the resignification of slowness as a life-enhancing quality has to be understood as a response to the acceleration society. As we saw in chapter 2, the experience of slowness emerged as a positive value in response to the velocity of newly emerging machines from the mid-nineteenth century. The shock induced by the speed of railway travel, for instance, provoked a renewed perspective on earlier modes of travel, such that walking could be redefined as offering leisure and heightened sensory pleasure. Slowness became recognized as a desirable or virtuous quality when it became a *choice,* rather than the only option, and when speed could be associated with negative characteristics such as alienation, stress, or desensitization. It is in this context that slowness could become the basis of formulating a critique of modernity: "speed *created* slowness, as it were."[17]

The Slow Food movement is worth considering briefly for the manner in which it treats time. Founded by the Italian food writer Carlo Petrini in response to the prospect of a McDonald's restaurant opening in the heart of Rome, the movement focuses on the contrast between "slow" and "fast" food in an attempt to demarcate a time, as much as a practice, distinct from the pace and pressures of work. It rejects globalized technologies, as well as the homogeneity and corporate greed associated with fast food production, and instead emphasizes the values associated with enjoyment, taste, authenticity, connectedness, tranquility, and community. Patterns of conviviality centered on eating slowly are celebrated as symbolic of making time for the important things in life. The assumption is that embedded within the time-honored practices of food preparation, rest, and hospitality lies a knowledge of mindful living.

The diversity of the movement makes it difficult to pin down the politics of slowness. In general, the Leftist origins of many participants mitigate conservative nostalgia for a lost organic community. Even so, a binary conception of slow as pleasure and speed as enslavement permeates their philosophy, as expressed in their manifesto: "Our century, which began and has developed under the insignia of industrial civilization, first invented the machine and then took it as its life model. We are enslaved by speed and have all succumbed to the same insidious virus: Fast Life."[18] Notably, however, in place of the fast life, slow food offers not only the pleasure of the table—taste, flavors, regionalism, locatedness—but also international exchange. Slowness is conceived of both as an individual, private subjectivity of "self-artistry," and as a social and political strategy for the betterment of society.

The shortcomings of the Slow Food movement are well rehearsed.[19] To mention a few: there is a tension between slowness as a realm of the elite (mostly Western) individual and the need to share the pleasures of slowness equitably; the need for an international movement that is itself opposed to globalization; a reliance on old discourses of pastoral versus high-density city life; and whether slow conviviality rests on a traditional domestic division of labor. More broadly, a social movement rooted in a politics of consumption cannot fully engage with the inequalities in time sovereignty that result from money, status, and power.

Shortcomings aside, such movements do open up a political space for questioning our obsession with speed as a virtue in and of itself. In defiance of the dominant time regime, a collective culture of deceleration would foster an alternative consumption of time—not just in the sense of more time, but more meaningful, deliberate and pleasurable time. As Wendy Parkins and Geoffrey Craig conclude, "The conscious cultivation of slowness may be a salutary reminder of how our rhythms and routines have the potential to either challenge or perpetuate the disaffection of everyday life."[20] By placing a fluid and dynamic understanding of time at the core of its philosophy, the Slow Food movement provides an interesting case study for examining how a more time-enriched lifestyle might be narrated.

As we move toward the final thoughts of this book, let me remind you why I would be reluctant to embrace slow living. Firstly, we cannot in fact choose between fast and slow, technology and nature. These dualisms exist and only acquire meaning in relation to each other. A fast/slow dichotomy cannot hope to capture the simultaneous coexistence of multiple temporalities that characterizes the experience of modernity. Once we recognize this, we can begin to reimagine hybrid sociomaterial assemblages or networks for enacting different times in an intensely technological world.

It follows, second, that a wholesale dismissal of globalization and digitalization as necessarily spurring on acceleration is misconceived. Even eating at a McDonald's restaurant can, in some contexts, be recuperated as a positive locale for slow eating, conviviality, and leisure. High-tech devices and systems are also great sources of pleasure and creativity. The positive possibilities for new kinds of time they generate should not be denied. Indeed, making more of time, preserving slow zones, actually *requires* more technological innovation.

New Technologies for Emergent Times

To argue, as I have in this book, that there is no temporal logic inherent in digital technologies is not to claim that technology is neutral or that the technical properties of objects do not matter. A sociotechnical lens on the sociology of time lays bare the highly specific materialities that make up the global network society. It demonstrates that the design and capabilities of the apparatus that become available to us, the architecture of infrastructures, have huge consequences. This has never been more so than now when every aspect of our lives is touched by information and communication technologies.

That some technologies are favored and developed while others languish neglected operationalizes the world in particular ways, obliging us to live accordingly. It closes off some options while opening up others. Understanding this dynamic is at the core of my thesis. To quote from Susan Douglas's social history of the radio: "Machines, of course, do not make history by themselves. But some kinds of machines help make different kinds of histories and different kinds of people than others."[21] We build our present and dream our future with and through tools and techniques and these visions are symbiotically reflected in them. So what forces are shaping technology today and what visions of the future are we being offered by the engineers of Silicon Valley?

What were once the stuff of science fiction are presented in daily media as if their realization is just around the corner. This near future is one peopled by robots and posthuman subjects with brains, bodies, and clothing enhanced by technoscience. According to the geeks of Silicon Valley, we are on evolutionary path to the next stage of a morphed cyborgian existence. Domestic robots feature prominently in these futuristic discourses. Electronic gadgets will clean floors, wash windows, scrub gutters, and even prepare healthy meals. Herb (the home exploring robot butler) developed by Carnegie Mellon University's Robotics Institute, for example, with two arms and a head-like box housing cameras and sensors, even speaks like a butler should, declaring in a Jeeves-like English accent, "I was designed to help people with household duties. One day I will help humans."[22] (This figure is especially appealing given the concurrent return of the real butler or manservant for the super-rich.)

In *The New Digital Age*, according to Google's Eric Schmidt and Jared Cohen, you will be roused by the aroma of freshly brewed coffee, with room temperature, humidity, music, and lighting all operating automatically, a gentle back massage administered by your high-tech bed that also guarantees a good night's sleep by measuring your REM cycle.[23] Your seamlessly interchangeable devices, some wearable, are all lightweight and incredibly fast and powerful. To paraphrase: the resulting gains in efficiency and productivity will be profound. By relying on these integrated systems, we'll be able to use our time more effectively each day—whether that means having a "deep think," spending more time preparing for an important presentation or guaranteeing that a parent can attend his or her child's football match without distraction. And, of course, the self-driving car will deliver you to work while you work!

These prospective technological scenarios seem inexhaustible and figure ever more powerfully in our culture. Time was when a new consumer durable had to be marketed with conventional accompanying advertisements, while now the latest versions of smartphones are newsworthy in themselves. They are not only reported in the financial pages but are headline news framing and representing tales of individual empowerment. The "i" word is practically an intoxicant. The topic of technology is in fact by far the best predictor of the popularity of news items on Twitter.[24]

At one level, all this speculative hyperbole has an obvious function: these (mostly) guys are promoting a benign future in which their own products feature heavily. They are transparently selling a particular vision in which technology will solve all our ills, including the time crunch. As I write, Big Data is the tool du jour for tech-savvy companies: the irresistible technological fix or answer to all social problems.[25] This idea, that we live in a technocracy in which technical rationality both defines political problems and provides the solution, has a long lineage. Frankfurt School writers like Herbert Marcuse and Jürgen Habermas were wise to it before the dawn of the computer.[26] Such depictions of the "proximate future" are far from innocent. They are being mobilized as a resource to influence the direction of sociotechnical innovation in the present.[27] Some thought-provoking sociology is exploring how patterns of hope, promise, and hype—the "dynamics of expectations"—are constructed, and the performative role they play in actually shaping research agendas.

Perhaps less obvious than the marketing ploys of the techno-evangelists is the extent to which speed itself has become the ultimate rationale for technical innovation. This in turn purveys a distorted model of the relationalities between time, technology and social change. Technologies change all the time, but this does not mean that technical change is always inventive.

In *The Mantra of Efficiency*, the historian Jennifer Karns Alexander traces how the modern orthodoxy that "all things should act efficiently" became dominant in Western culture. "Good" technological design is efficient, it is about making things work, effecting control over situations and events. This is "particularly apparent in the contemporary emphasis on quantifiable productivity and associated fears of waste, especially the waste of time."[28] In other words, technological inventiveness is associated with making us more efficient in the sense of being economical with time.

This instrumental philosophy of maximizing efficiency is at the heart of engineering. According to this logic, automation is the perfect solution because human "interference" is a potential source of error and should be eliminated. The latest, fastest, most automated systems appear as objectively the best, rather than as the congealed product of particular localized choices, histories, ideas, technical instruments, and materials.[29]

Take something we rarely think about, searching the web. We interface with IT apparatus as if it was immaterial, screens providing neutral, value-free information. The speed of Google's search engine so enthralls us that we seldom reflect on the fact that it favors some content over others. Nor are we alert to the way that entering a search term into Google triggers an instant auction that determines the order in which advertisements appear. This is especially true for the all-important first page of search results. What matters here, the way power is exercised, is in what is excluded—the vast number of relevant websites that we do not see.

It used to be the case (until Google changed the results) that a search for the phrase, "she invented," would return the query, "Did you mean "he invented?"." As Google explained, this "correctly" reflects past searches, in that over the entire corpus of the web the word "invented" is preceded by "he" much more often than "she." Google's algorithm recognized this—and presumed it meant the first search query was merely a typographical error. Thus the conventional wisdom that the world's greatest inventors are male presents itself as factual. We could have more diverse search engine technology, running different algorithms, instead of the standardization of the key filter for most web users.[30] This would be much slower. But might it be more efficient in the sense of acknowledging the difference between data, information and knowledge? Might the articles that a computer program quickly tells you are "popular" on the Internet not necessarily be the same as those that are actually worth reading?

There is growing recognition that software algorithms are not impartial.[31] Different software embeds different philosophies, and these philosophies, as they become ubiquitous, become invisible. Code is written by people, it contains inferences and assumptions and embodies certain human values and biases. Indeed, the surface structure of flexibility and freedom, heralded by computing, is undergirded by a rigid infrastructure of legal and economics protocols. Software systems shape output—computer-generated predictions, recommendations, and simulations. What appears to be efficient, the constant upgrading of computer software and hardware packages, "are instances of a restrictive strategy, locking users into existing configurations producing enforced obsolescence, reproducing the contours of the existing technological zone in a trivially 'new' form."[32] The flip side of accelerated novelty production, the continual simulation of the new, is a mounting pile of trash.

But this need not be the case. Jonathan Sterne, for example, imagines a company that took its time developing a computer that could last, could be easily updated, repaired, and upgraded, was easy to learn and use, worked well with other platforms, and that was less environmentally hazardous when it did finally decompose.[33] This would be a "convivial" tool in Ivan Illich's sense: ease of use, flexibility in implementation, harmony with the environment, and ease of integration into truly democratic forms of social life. Instead, computers are designed to become obsolete after a short period of use. Within the occupational ideology of computer engineering, Moore's law is "less of a law of computer evolution than it is a fantasy the industry wishes to uphold," as a high rate of machine turnover drives exponential profit.

To be sure, we have been all too ready to conflate the speed of technical innovation with inventiveness. On the contrary, rapid technological change can actually be conservative, maintaining or solidifying existing social arrangements. Its very speed may occur in order to block and stifle the possibility of alternative trajectories. As other STS scholars argue, inventiveness is not about the novelty of artifacts in themselves, but about the degree to which they are "aligned with inventive ways of thinking and doing and configuring and reconfiguring relations with other actors."[34] Genuine inventiveness, then, can occur when the pace of technological change is slow, or in places and at times least expected.

If inventiveness is about challenging our common-sense ways of doing things, questioning the assumptions that permeate our political discourse, and creating new possibilities for the present, we leave engineering to the engineers at our peril. I have written elsewhere about the culture of engineering and computing, where "the masculine workplace culture of passionate virtuosity, typified by hacker-style work, epitomizes a world of mastery, individualism and non-sensuality."[35] Being in an intimate relationship with a computer can be both a substitute for, and a refuge from, the much more uncertain and messy relationships that characterize social life. It is an environment that thwarts the imaginations of technology designers, ignoring the needs of those who do not fit or conform to their own paradigm of normality.[36] One might venture that this mindset is ever more influential in our digital age in which the world's richest companies are predominantly engineering companies: Microsoft, Apple, Google, Facebook, and Twitter.

My point here is not about the predisposition of individual engineers so much as the institutional culture they inhabit. Much technoscientific innovation originates in either the military or in corporate business environments, where the expertise and ingenuity that is valued is directed toward tackling certain kinds of soluble problems.

Take a radical innovation for saving time, such as Google's driverless car. That a car can drive itself without crashing is a remarkable achievement of mapping software. It is economical with time, in that you can get on with work while being driven and don't need to employ a chauffeur to achieve this efficiency. However, as Evgeny Morozov rightly points out, there may be unintended consequences: "Would self-driving cars result in inferior public transport as more people took up driving? Would it lead to even greater suburban sprawl as, now they no longer had to drive, people could do email during their commute and thus would tolerate spending more time in the car?"[37] One might add that little thought has been given to gendered patterns of travel, the intricate map of itineraries that thread through the daily lives of mothers. Car travel is not wholly instrumental. It may be an important time and space for parent-child relationships, such as on the daily school run. As the driverless car becomes more reliable, parents will surely succumb to the temptation to send their children off in it while they get on with other tasks.

But in relation to saving time, the self-driving car presents a narrow model of change, even in terms of transport. The car is more than a machine for mobility, it is a sociotechnical system that locks people into certain social habits and practices. Shifting these requires innovating in the economic, political, and social arrangements that embed them. In fact, cars are in decline in the West, and travel activity has reached a plateau. Many forecast alternative scenarios of electric cars based not on individual ownership but on access.[38] They cite new vehicle-sharing systems, the increasing interest of car manufacturers in experimenting with pay-as-you-go schemes, and the mass development of electric bikes in China. What is at stake here is commuting time. But perhaps long distance travel may become less necessary with hypermobile network communications while refinements to less sophisticated technologies, such as the bicycle, increase their appeal. In other words, a more imaginative combination of old and new technologies together with different ownership arrangements might release more time than automating the car.

Enthusiasm for the smart kitchen also betrays a belief that automation automatically saves time, regardless of the context. The idea of a kitchen run by super-intelligent machines, like Herb the robot butler, is not radical at all. As we saw in chapter 5, there are limits to how smart a kitchen can be if the home is conceived in terms akin to a rational, orderly, well-functioning machine. Time is treated as a highly individualized activity rather than as an aspect of shared, socially organized activities that are themselves shaped both by institutions and physical infrastructures. A far more inventive design frame would think outside the kitchen box. It would see beyond the private single-family household to encompass different social arrangements. If the aim were to economize on housework time, it could reorganize the sexual division of domestic labor. It might even collectivize housework in the way envisioned by American feminists like Charlotte Perkins Gilman in the late nineteenth century.[39]

The problem is that ultra high-tech conceptions of conserving personal time have ramifications. They materially shape not only our machines but also our cultural frames, tropes, and metaphors for understanding ourselves. How can we possibly refuse the ensemble of objects that define our world as given, if we are constantly narrated as subjects who are programmed, "hard-wired," or coded to process information in particular ways? As Lucy Suchman has shown, grand projects in humanoid robotics and artificial intelligence perversely limit the potential for altering our cultural imaginaries of the human. They make it much more difficult to locate "the conditions for action and possibilities for intervention in the specificities of mundane sociomaterial assemblages."[40]

To my mind, such projects also reflect the marginalization of women from technoscientific work. (Even the famous MIT Media Lab has a male milieu, as only 20 percent of the faculty are women.[41]) Mine is not an essentialist argument about innate female values but part of a more general proposition about opening up the processes of technological research and development to a wider range of societal groups and interests. Time is calibrated by power, and therefore promoting diversity in design would produce entities and interfaces more attuned to those who do not fit the mold of methods oriented to speed. The way ICTs are interpreted and used depends on the tapestry of social relations woven by age, gender, race, class, and other axes of inequality. We should not subsume all the often slow and discordant rhythms of everyday life to the standardized clock time of our technoculture.

The digital is not a black box, a magic thing that is going to fulfill a vision of the future. It is coproduced with society and mirrors the boundaries of our imagination. Rather like the ways in which people represent themselves on social media, the initial radical promises of cyberspace as a disembodied zone of freedom is belied by clichéd and deeply regressive visual and textual representations. It is new technology reiterating old narratives. What presents itself as novelty and change is often a more concrete expression of the ongoing limits to, and stasis in, our collective social and political aspirations—as these are continually enacted in the distorted distributions of labor, time, power, and other resources.

* * *

The contemporary imperative of speed is as much a cultural artifact as a material one. We are at a pivotal juncture in the evolution of technology, yet industrial regimes still set the contours of our lives. These traditional timescapes and domains of activity have been subsumed within networked 24/7 digital temporalities. The question posed by this book is whether acceleration is an adequate trope for understanding our emerging relationship to time.

Rather than being endemically pressed for time, perhaps we are confused about what time we are living in. Part of the problem may be that the categories of speed and acceleration, and their association with progress, productivity and efficiency, do not provide us with the appropriate language to formulate fresh ideas about how we might leverage the digital infrastructure.

Electronic technologies are integral to our experience of space, time, communication, and consciousness, crystallizing new ways of being, knowing, and doing. They as much reflect our high-speed culture as shape it. If technologies are sites of practice, then sociotechnical orders are not predetermined but are the result of humans and nonhumans coming together to constitute society. The latest technologies can, then, be recruited as a resource in our quest for discretionary time.

Too often, however, critical reflections on the impact of digital devices are framed negatively, as if we are victims of a "crisis" that needs correction.[42] Such readings make it difficult to formulate an alternative politics of time (and in particular a gendered time politics) which cannot be separated from either the emergence of digitalization, or its entanglement with the shifting temporalities of social life.

There is a disjunction between the cultural allure of speed and the common experience of always feeling rushed, but this can be a source of creative tension. Smart, fast technologies provide an unparalleled opportunity for realizing a more humane and just society, only we need to keep in mind that busyness is not a function of gadgetry but of the priorities and parameters we ourselves set. Now is the moment to contest the euphorics of speed, and the technological impulse to achieve it, harnessing our inventiveness to take control of our time more of the time.

Notes

Introduction

1 Helga Nowotny, *Time: The Modern and Postmodern Experience* (Cambridge: Polity Press, 2005), 18.
2 Gordon E. Moore, "Cramming More Components onto Integrated Circuits," *Electronics* 38, no. 8 (1965): 114–17, described how the number of transistors that can be placed on a circuit doubles every two years.
3 I do not intend to go into the vast literature on the philosophy of time. See, for example, chapter 7 of *Time Matters: On Theory and Method* (Chicago: University of Chicago Press, 2001), where Andrew Abbott draws on Henri Bergson, George Herbert Mead, and Alfred North Whitehead to elaborate a theory of temporality as social, processual, and relational. To paraphrase, what matters is not how fast social change is occurring or that new communication technologies make interaction faster, but rather how fast these things occur *relative* to other things as they used to be.
4 Donald MacKenzie, Daniel Beunza, Yuval Milo, and Juan Pablo Pardo-Guerra, "Drilling through the Allegheny Mountains: Liquidity, Materiality and High-Frequency Trading," *Journal of Cultural Economy* 5, no. 3(2012): 279–96. For example, a firm that did not use it would have feared that its price quotes would become "stale" (that is, they would no longer reflect price changes in the wider market), and that those stale quotes would be "picked off" by trading firms using the new, faster, cable.
5 I refer to *social shaping* here simply to introduce the idea that technology is socially shaped, but society is technically shaped, too. See Donald MacKenzie and Judy Wajcman, eds., *The Social Shaping of Technology*, 2nd ed. (Milton Keynes, UK: Open University Press, 1999). A fuller account of science and technology studies can be found in the second half of chapter 1.
6 See John Robinson and Geoffrey Godbey, "Busyness as Usual," *Social Re-*

search 72, no. 2 (2005): 407–26. Details of all these studies can be found in chapter 3.

7 Jonathan Gershuny and Kimberly Fisher calculate the figure of five hundred minutes as consistent across the developed world from 1961 to 2006. See "Exploit and Industry: Why Work Time Will Not Disappear for Our Grandchildren" (paper presented at The Value of Time: Addressing Social Inequalities: 35th IATUR Conference on Time Use Research, Rio de Janeiro, Brazil, August 7–10, 2013).

8 See Manuel Castells, *The Rise of the Network Society* (Oxford: Blackwell, 1996), and John Urry, *Sociology Beyond Societies: Mobilities for the Twenty-First Century* (London: Routledge, 2000) respectively, detailed in chapter 1.

Chapter One

1 Philipp Hildebrand, vice-chairman of BlackRock, goes on to say that he has meditated for seven years and that "in the financial world, it is a must" ("Zen and the Art of Management," *Financial Times*, September 17, 2013).

2 Manuel Castells, *The Rise of the Network Society* (Oxford: Blackwell, 1996), uses the term "timeless time"; David Harvey, *The Condition of Postmodernity* (Oxford: Blackwell, 1990), uses "time-space compression"; Anthony Giddens, *The Consequences of Modernity* (Cambridge: Polity Press, 1990), uses "time-space distanciation"; John Urry, *Sociology Beyond Societies: Mobilities for the Twenty-First Century* (London: Routledge, 2000), uses "instantaneous time"; Paul Virilio, *Speed and Politics: Second Edition* (New York: Semiotext[e], 1986), uses "chronoscopic time"; Robert Hassan, *Empires of Speed: Time and the Acceleration of Politics and Society* (Leiden: Brill Academic Publishers, 2009), uses "network time"; and Michel Maffesoli, *L'instant eternal* (Paris: La Table Ronde, 2003), uses "pointillist time." See also Helga Nowotny, *Time: The Modern and Postmodern Experience* (Cambridge: Polity Press, 2005), and various writings by Barbara Adam, such as *Timewatch: The Social Analysis of Time* (Cambridge: Polity Press, 1995); "Reflexive Modernization Temporalized," *Theory, Culture & Society* 20, no. 2 (2003): 59–78; *Time* (Cambridge: Polity Press, 2004).

3 Hartmut Rosa has developed his thesis that social acceleration is the constitutive trait of post- or late modernity in a range of impressive publications, such as "Social Acceleration: Ethical and Political Consequences of a Desynchronized High-Speed Society," *Constellations* 10, no. 1 (2003): 3–33 (see esp. p. 28); *Social Acceleration: A New Theory of Modernity* (New York: Columbia University Press, 2013).

4 Rosa, "Social Acceleration," 10.

5 Harvey, *The Condition of Postmodernity*, 240. See also Marshall McLuhan, *The Gutenberg Galaxy: The Making of the Typographic Man* (Toronto: University of Toronto Press, 1962).

6 Adam, "Reflexive Modernization Temporalized," 67.

7 Donald Mackenzie, "How to Make Money in Microseconds," *London Review of Books*, May 19, 2011, 16–18.

8 John Stephens, "World's Fastest Internet Speed: 186 Gbps Data Transfer Sets New Record," *Huffington Post*, June 29, 2012. http://www.huffingtonpost.com/2011/12/16/worlds-fastest-internet_n_1154065.html.

9 Carmen Leccardi, "Resisting 'Acceleration Society,'" *Constellations* 10, no. 1 (2003): 37.

10 Manuel Castells's wide-ranging analysis has been the subject of much critical commentary. See, for example, Frank Webster and Basil Dimitriou, eds., *Manuel Castells* (London: Sage, 2003).

11 Manuel Castells, *The Rise of the Network Society*, 2nd ed. (Malden, MA: Blackwell, 2010), xii.

12 Ibid., 467.

13 Urry, *Sociology Beyond Societies*, 129, 126.

14 Barry Brown and Kenton O'Hara, "Place as a Practical Concern of Mobile Workers," *Environment and Planning A* 35, no. 9 (2003): 1565–87.

15 Gina Neff, *Venture Labor: Work and the Burden of Risk in Innovation Industries* (Cambridge, MA: MIT Press, 2012).

16 Worldwide, digital warehouses use about thirty billion watts of electricity, roughly equivalent to the output of thirty nuclear power plants, and 90 percent of it is wasted. See "Power, Pollution and the Internet," *New York Times*, September 30, 2012.

17 For example, the trading of shares, foreign exchange, futures, and options is quite distinctive, and this is not simply to do with the nature of these financial products. The balance of power between exchanges (such as the New York Stock Exchange), big banks, and financial regulators has shaped these markets in significantly different ways. See MacKenzie et al., "Drilling through the Allegheny Mountains: Liquidity, Materiality and High-Frequency Trading," *Journal of Cultural Economy* 5, no. 3 (2012): 279–96.

18 See Andrew Blum's popular account of the material infrastructure of the Internet, *Tubes: Behind the Scenes at the Internet* (London: Viking, 2012), 9.

19 Carolyn Marvin, *When Old Technologies Were New: Thinking About Electric Communication in the Late Nineteenth Century* (New York: Oxford University Press, 1988), 4. Within media and communication studies, the same critique is often referred to as "media centrism." For a discussion of the contrasting approaches of media studies and STS, see Judy Wajcman and Paul Jones, "Border Communication: Media Sociology and STS," *Media, Culture & Society* 34, no. 6 (2012): 673–90.

20 See Nigel Thrift, "New Urban Eras and Old Technological Fears: Reconfiguring the Goodwill of Electronic Things," *Urban Studies* 33, no. 8 (1996): 1467.

21. See Donna Haraway, *Modest—Witness@Second—Millennium.Female-Man—Meets—OncoMouse:Feminism and Technoscience* (New York: Routledge, 1997).

22 Paul Virilio, "Speed-Space: Interview with Chris Dercon," in *Virilio Live: Selected Interviews*, ed. John Armitage (London: Sage, 2001), 70. See also *The*

Art of the Motor (Minneapolis: University of Minnesota Press, 1995); *Open Sky* (London: Verso, 1997); and *The Futurism of the Instant: Stop-Eject* (Cambridge: Polity Press, 2010).

23 Bertrand Richard, in Paul Virilio, *The Administration of Fear* (Los Angeles: Semiotext[e], 2012), 7.

24 See Steve Redhead, *Paul Virilio: Theorist for an Accelerated Culture* (Toronto: University of Toronto Press, 2004).

25 The acceleration of transport also has dire consequences for both energy use and pollution, and Virilio despairs about our failure to recognize the fundamental disjuncture between industrial time and the timescapes of environmental hazards. See Barbara Adam, *Timescapes of Modernity: The Environment and Invisible Hazards* (London: Routledge, 1998), for an eloquent account of the temporal aspects or "timescapes" of environmental hazards.

26 At such points, Virilio does move beyond seeing speed in purely technical terms, taking into account the socioeconomic relations within which it is embedded. However, this is less evident in his discussion of the effects of transmission and transplantation.

27 For Virilio, technological evolutionism always means *military*-technological evolutionism (Stefan Breuer, "The Nihilism of Speed: On the Work of Paul Virilio," in *High-Speed Society: Social Acceleration, Power, and Modernity*, ed. Hartmut Rosa and William Scheuerman, 215–41 [University Park, PA: Penn State University Press, 2009]).

28 Ibid., 227.

29 Adam, "Reflexive Modernization Temporalized," 71.

30 Jonathan Crary, *24/7: Late Capitalism and the Ends of Sleep* (London: Verso, 2013), 11. See also *Thesis Eleven* 118, no. 1 (2013), on the theme of technology and time in liquid modernity.

31 Robert Hassan, *Chronoscopic Society: Globalization, Time and Knowledge in the Network Economy* (New York: Peter Lang, 2003), 236.

32 Robert Hassan, "Social Acceleration and the Network Effect: A Defence of Social 'Science Fiction' and Network Determinism," *British Journal of Sociology* 61, no. 2 (2010): 368.

33 Donna Haraway, "A Cyborg Manifesto: Science, Technology, and Socialist-Feminism in the Late Twentieth Century," in *Simians, Cyborgs and Women: The Reinvention of Nature*, ed. Donna Haraway (New York: Routledge, 1991).

34 William Connolly, "Speed, Concentric Cultures, and Cosmopolitanism," in Rosa and Scheuerman, *High-Speed Society*, 263.

35 Donald MacKenzie and Judy Wajcman, eds., *The Social Shaping of Technology*, 2nd ed. (Milton Keynes, UK: Open University Press, 1999), 4.

36 This is in no way to deny the rich theoretical tradition of media studies, and the increasing crossover interest in materiality in those fields. See Wajcman and Jones, "Border Communication"; Tarleton Gillespie, Pablo Boczkowski, and Kirsten Foot, eds., *Media Technologies: Essays on Communication, Materiality, and Society* (Cambridge, MA: MIT Press, 2014). There is

also an affinity with the material cultural perspective associated with cultural anthropology, see Daniel Miller, *Material Culture and Mass Consumption* (Oxford: Blackwell, 1991).

37 STS is a very broad field with distinct theoretical positions that have often rubbed against one another quite sharply. See, for example, some of the debates about the nature of STS in the *Social Studies of Science* journal. There is no space to go into the details of these debates here—and anyway it's the commonality of the approaches that I want to emphasize. A broad sample of the research can be found in journals such as *Science, Technology, & Human Values* and *Science as Culture*; the book series *Inside Technology* (MIT Press); and Edward Hackett, Olga Amsterdamska, Michael Lynch, and Judy Wajcman, eds., *The Handbook of Science and Technology Studies*, 3rd ed. (Cambridge, MA: MIT Press, 2008).

38 Ruth Schwartz Cowan, "From Virginia Dare to Virginia Slims: Women and Invention in America," *Technology and Culture* 20, no. 1 (1979): 52.

39 William Dutton, *Society on the Line: Information Politics in the Digital Age* (Oxford: Oxford University Press, 1999); and William Dutton, ed., *The Oxford Handbook of Internet Studies* (Oxford: Oxford University Press, 2013).

40 See STS studies, such as Wiebe Bijker and John Law, *Shaping Technology/Building Society: Studies in Sociotechnical Change* (Cambridge, MA: MIT Press, 1992); MacKenzie and Wajcman, *The Social Shaping of Technology*; Judy Wajcman, *TechnoFeminism* (Cambridge: Polity Press, 2004).

41 See Michel Callon, "Some Elements of a Sociology of Translation: Domestication of the Scallops and the Fishermen of Saint Brieuc Bay," in *Power, Action and Belief: A New Sociology of Knowledge?*, ed. John Law (London: Routledge, 1986); Madeleine Akrich and Bruno Latour, "A Summary of a Convenient Vocabulary for the Semiotics of Human and Nonhuman Assemblies," in Bijker and Law, *Shaping Technology/Building Society*, 259–64; Bruno Latour, *Aramis, or The Love of Technology* (Cambridge, MA: Harvard University Press, 1996); Bruno Latour, *Reassembling the Social: An Introduction to Actor–Network Theory* (Oxford: Oxford University Press, 2005); John Law and John Hassard, eds., *Actor Network Theory and After* (Oxford: Blackwell, 1999).

42 Some scholars have now moved from a focus on interaction to an interest in "intra-action" in order to get a better grasp of the way that distinct things like humans and machines get formed through their interactions in the first place. See Karen Barad, "Posthumanist Performativity: Toward an Understanding of How Matter Comes to Matter," *Signs: Journal of Women in Culture and Society* 28, no. 3 (2003): 801–31; and Lucy Suchman, *Human-Machine Reconfigurations: Plans and Situated Actions*, 2nd ed. (New York: Cambridge University Press, 2007). STS scholars have deployed the concept of "performativity" to encapsulate the idea that the properties of and boundaries between humans and technologies are not fixed in advance but enacted in recurrent activities. This concept is a familiar one in feminist theory through Judith Butler's argument that gender is a per-

formance, *Gender Trouble: Feminism and the Subversion of Identity* (New York: Routledge, 1990).

43 Hartmut Rosa, "Social Acceleration: Ethical and Political Consequences of a Desynchronized High-Speed Society" in Rosa and Scheuerman, *High-Speed Society*, 91.

Chapter Two

1 Edward P. Thompson, "Time, Work-Discipline and Industrial Capitalism," *Past & Present* 38, no. 1 (1967): 56–97.

2 Jeremy Rifkin, *Time Wars* (New York: Henry Holt, 1987), 3–4.

3 Charles Dickens, *Hard Times* (1854; London: Penguin, 2003), 35.

4 Barbara Adam, "Reflexive Modernization Temporalized," *Theory, Culture & Society* 20, no. 2 (2003): 63. See Barbara Adam, *Time* (Cambridge: Polity Press, 2004), chapter 6, for a discussion of the four Cs of industrial time and her overarching concept for the cluster, or *timescape*, of industrial societies' temporal relations.

5 Helga Nowotny, *Time: The Modern and Postmodern Experience* (Cambridge: Polity Press, 1994), 84.

6 See, for example, Anthony Giddens, *A Contemporary Critique of Historical Materialism* (London: Macmillan, 1981); David Harvey, *The Condition of Postmodernity* (Oxford: Blackwell, 1990).

7 Karl Marx, *Capital*, vol. 1 (New York: International Publishers, 1967), 233; Harvey, *The Condition of Postmodernity*, 230.

8 Thompson, "Time, Work-Discipline and Industrial Capitalism," 82–86.

9 Allen Bluedorn, *The Human Organization of Time: Temporal Realities and Experience* (Stanford, CA: Stanford University Press, 2002), 92. See also Michel Foucault, *Discipline and Punish: The Birth of the Prison* (London: Penguin, 1985).

10 Paul Glennie and Nigel Thrift, *Shaping the Day: A History of Timekeeping in England and Wales 1300–1800* (Oxford: Oxford University Press, 2011).

11 Ibid., 231.

12 Ibid., 13.

13 Indeed, Hannah Gay, "Clock Synchrony, Time Distribution and Electrical Timekeeping in Britain 1880–1925," *Past & Present* 181, no. 1 (2003): 107–40, argues that Thompson exaggerates the causal role of clocks in establishing factory discipline, since exact and binding clocks only became widespread at the end of the nineteenth century.

14 Ibid., 140.

15 Stephen Kern, *The Culture of Time and Space 1880–1919* (London: Weidenfeld and Nicolson, 1983).

16 Tom Standage, *The Victorian Internet: The Remarkable Story of the Telegraph and the Nineteenth Century's On-Line Pioneers* (London: Weidenfeld and Nicolson, 1998).

17 For an excellent discussion of the different meanings and usage of the terms modern and modernism, see Paul Jones and David Holmes, *Key Concepts in Media and Communications* (London: Sage, 2011).
18 John Tomlinson, *The Culture of Speed: The Coming of Immediacy* (London: Sage, 2007), 20–23.
19 Ibid., 22, 23.
20 James Carey, "Technology and Ideology: The Case of the Telegraph," *Prospects* 8 (1983): 304.
21 Lord Salisbury, quoted in Gay, "Clock Synchrony, Time Distribution and Electrical Timekeeping in Britain 1880–1925," 127.
22 Anthony Giddens, *The Consequences of Modernity* (Cambridge: Polity Press, 1990), 53, 16–17.
23 For an excellent account of the constitutive role of communications in the processes of modernity, see Graham Murdock, "Communications and the Constitution of Modernity," *Media, Culture & Society* 15, no. 4 (1993): 521–39.
24 Carey, "Technology and Ideology," 319. As Carey also notes, these were the very same conditions that underlay Marx's analysis of the commodity fetish.
25 H. G. Wells, *The Time Machine* (London: William Heinemann, 1895); Albert Einstein's 1905 papers, in John Stachel, ed., *Einstein's Miraculous Year: Five Papers that Changed the Face of Physics* (Princeton, NJ: Princeton University Press, 2005).
26 Marshal Berman, *All That Is Solid Melts into Air: The Experience of Modernity* (London: Verso, 1983), 15. See also Harvey, *The Condition of Postmodernity*, part 3.
27 Le Corbusier, "A Contemporary City," in *The City Reader*, 4th ed., ed. Richard LeGates and Frederic Stout (London: Routledge, 2007), 330.
28 Tomlinson, *The Culture of Speed*, 47.
29 Lucy Suchman, "Affiliative Objects," *Organization* 12, no. 3 (2005): 379–99; Sherry Turkle, ed., *Evocative Objects: Things We Think With* (Cambridge, MA: MIT Press, 2007); Judy Wajcman, *Feminism Confronts Technology* (Cambridge: Polity Press, 1991).
30 Tomlinson, *The Culture of Speed*, 65.
31 Norbert Elias, "Technization and Civilization," *Theory, Culture & Society* 12, no. 3 (1995): 7–42.
32 Mike Featherstone, "Automobilities: An Introduction," *Theory, Culture & Society* 21, no. 4/5 (2004): 3. See this special issue for recent sociological approaches, such as treating the driving body as "a set of social practices, embodied dispositions, and physical affordances," and as a sociotechnical "hybrid" or "assemblage."
33 Saskia Sassen, *Cities in a World Economy* (Los Angeles: Sage, 2012).
34 Georg Simmel, *The Philosophy of Money* (London: Routledge, 1978), 450.
35 Ibid., 506.
36 Georg Simmel, "The Metropolis and Mental Life," in *Simmel on Culture:*

Selected Writings, ed. David Frisby and Mike Featherstone (London: Sage, 1997), 175 (original emphasis). See also Lawrence Scaff, "The Mind of the Modernist: Simmel on Time," *Time & Society* 14, no. 1 (2005): 5–23.

37 Manfred Garhammer, "Pace of Life and Enjoyment in Life," *Journal of Happiness Studies* 3, no. 3 (2002): 248 (original emphasis).

38 Scaff, "The Mind of the Modernist," 18. See also Frisby and Featherstone, *Simmel on Culture*.

39 *Ford Times*, 1913. Tim Leunig and Hans-Joachim Voth, "Spinning Welfare: The Gains from Process Innovation in Cotton and Car Production" (discussion paper no. 1050, London: LSE, Centre for Economic Performance, May 2011).

40 Virginia Scharff, *Taking the Wheel: Women and the Coming of the Motor Age* (Albuquerque: University of New Mexico Press, 1992).

41 Paul Gilroy, *Darker than Blue: On the Moral Economies of Black Atlantic Culture* (Cambridge, MA: Harvard University Press, 2010), 34, notes that the same car that was for blacks part of their belated liberation from their US apartheid, for whites supplied a means of perpetuating racial segregation: "White flight from urban centers was not just accomplished by means of the automobile—it was premised on it."

42 John Urry, "The 'System' of Automobility," *Theory, Culture & Society* 21, no. 4/5 (2004): 25–39.

43 Transport for London, "Analysis of Cycling Potential: Policy Analysis Research Report" (London: TfL, 2010). http://www.tfl.gov.uk/assets/downloads/corporate/analysis-of-cycling-potential.pdf.

44 Robert Gordon, "Why Innovation Won't Save Us," *Wall Street Journal*, December 21, 2012.

45 As Raymond Williams, *Television: Technology and Cultural Form* (London: Fontana, 1974), 26, famously argued, the private car led to *mobile privatization*: "the two paradoxical yet deeply connected tendencies of modern urban industrial living: on the one-hand mobility, on the other hand the more apparently self-sufficient family home." I will pick up this theme again in chapter 6 where I discuss the portability of digital devices, such as the mobile phone.

46 *Lancet* 380, no. 9838 (July 18, 2012). See also Lyndall Strazdins et al., "Time Scarcity: Another Health Inequality?," *Environment and Planning A* 43, no. 3 (2011): 545–59. These Australian epidemiologists frame time as a health resource, for example, arguing that time pressure is the main reason people give for not taking exercise. As such, time scarcity generates health inequality. In the next chapter we will see that men and women's time use differs, including importantly, their leisure time.

47 John Urry, *Sociology Beyond Societies* (London: Routledge, 2000) and *Mobilities* (Cambridge: Polity Press, 2007).

48 See, for example, Zygmunt Bauman, *Liquid Modernity* (Cambridge: Polity Press, 2000).

49 David Morley, *Home Territories: Media, Mobility and Identity* (London: Routledge, 2000), 14.

50 Tim Cresswell, "Towards a Politics of Mobility," *Environment and Planning D: Society and Space* 28 (2010): 23.

51 Beverley Skeggs, *Class, Self, Culture* (London: Routledge, 2004), 60.

52 Janet Wolff, "The Invisible Flâneuse: Women and the Literature of Modernity," *Theory, Culture & Society* 2, no. 3 (1985): 37–46.

53 "World-wide simultaneity" is the term coined by Helga Nowotny, *Time*, and, as we shall see later, it has much in common with Tomlinson's "immediacy" (*The Culture of Speed*).

54 See also Mike Savage, "Against Epochalism: An Analysis of Conceptions of Change in British Sociology," *Cultural Sociology* 3, no. 2 (2009): 217–38.

Chapter Three

1 Joseph Stiglitz, Amartya Sen, and Jean-Paul Fitoussi, "Report by the Commission on the Measurement of Economic Performance and Social Progress" (2009): 12 (original emphasis). http://www.stiglitz-sen-fitoussi.fr.

2 See, for example, Richard Layard, *Happiness: Lessons From A New Science* (London: Penguin, 2011).

3 Robert Goodin, James Rice, Antti Parpo, and Lina Eriksson, *Discretionary Time: A New Measure of Freedom* (Cambridge: Cambridge University Press, 2008), 4. See also Michael Marmot's classic Whitehall study, which shows that lower-grade civil servants, with less control over their time, have higher mortality rates than those in senior ranks, *Status Syndrome: How Your Social Standing Directly Affects Your Health and Life Expectancy* (London: Bloomsbury, 2004).

4 Helga Nowotny, *Time: The Modern and Postmodern Experience* (Cambridge: Polity Press, 2005).

5 Juliet Schor, *The Overworked American: The Unexpected Decline of Leisure* (New York: Basic Books, 1991). See also Arlie Hochschild, *The Time Bind: When Work Becomes Home and Home Becomes Work* (New York: Metropolitan Press, 1997).

6 Suzanne Bianchi, John Robinson, and Melissa Milkie, *Changing Rhythms of American Family Life* (New York: Russell Sage, 2006).

7 Cynthia Fuchs Epstein and Arne Kalleberg, eds., *Fighting For Time: Shifting Boundaries of Work and Social Life* (New York: Russell Sage, 2004); Ellen Galinsky, *Ask the Children: What America's Children Really Think About Working Parents* (New York: William Morrow, 1999); Jonathan Gershuny, *Changing Times: Work and Leisure in Postindustrial Society* (New York: Oxford University Press, 2000); Jerry Jacobs and Kathleen Gerson, *The Time Divide: Work, Family and Gender Inequality* (Cambridge, MA: Harvard University Press, 2004).

8 John Robinson and Geoffrey Godbey, *Time for Life: The Surprising Ways Americans Use Their Time* (University Park: Penn State University Press, 1997), 231; Jonathan Gershuny, "Busyness as the Badge of Honor for the New Superordinate Working Class," *Social Research: An International Quarterly* 72, no. 1 (2005): 287–314; and John Robinson "Americans Less Rushed But No Happier: 1965–2010 Trends in Subjective Time and Happiness," *Social Indicators Research* 113 (2013): 1091–1104. Robinson's latest figures (from 2010) show an unexpected decline from 2004 in the proportion of those always rushed, and he speculates that this may reflect a simple adaption to an ever-faster pace of life.

9 Among time-use specialists, the term "free time" is defined as the remaining time after time spent in market and nonmarket work and meeting physiological needs (sleeping, eating, attending to personal hygiene, and grooming) is deducted. It represents the time potentially available for leisure pursuits.

10 Goodin et al., *Discretionary Time*, 73.

11 Robinson and Godbey, *Time for Life*; Gershuny, Busyness as the Badge of Honor, 288; Gershuny and Kimberly Fisher (2013), "Exploit and Industry: Why Work Time Will Not Disappear for Our Grandchildren," *Centre for Time Use Research*, Oxford University. http://www.timeuse.org/node/6934.

12 Paul Edwards and Judy Wajcman, *The Politics of Working Life* (Oxford: Oxford University Press, 2005).

13 Jacobs and Gerson, *The Time Divide*, 39. The long hours of low paid workers doing multiple jobs do not figure in these discussions!

14 Man Yee Kan, Oriel Sullivan and Jonathan Gershuny, "Gender convergence in domestic work: discerning the effects of interactional and institutional barriers from large-scale data," *Sociology*, 45, 2 (2011), 234–51; Harriet Presser, "Race-ethnic and gender differences in nonstandard work shifts," *Work and Occupations*, 30, 4 (2003), 412–39.

15 Kim Parker and Wendy Wang, *Modern Parenthood* (Pew Research Centre, 2013), 4.

16 Michael Bittman, "Parenting and Employment: What Time-Use Surveys Show," in *Family Time: The Social Organization of Care*, ed. Nancy Folbre and Michael Bittman (London: Routledge, 2004). See also Marybeth Mattingly and Suzzane Bianchi, "Gender Differences in the Quantity and Quality of Free Time: The US Experience," *Social Forces*, 81 (2003), 999–1029.

17 Karen Davies, *Women, Time and the Weaving of the Strands of Everyday Life* (Aldershot, UK: Avebury, 1990); Miriam Glucksmann, "'What a Difference a Day Makes': A Theoretical and Historical Exploration of Temporality and Gender," *Sociology* 32, no. 2 (1998): 239–58; Oriel Sullivan, "Time Waits for No (Wo)man: An Investigation of the Gendered Experience of Domestic Time," *Sociology* 31, no. 2 (1997): 221–39.

18 In the United States, only about two-thirds of children under five years old now live with married parents (biological, step, or adoptive), and this declines to about 55 percent for children aged fifteen to seventeen. There has

been a dramatic increase in divorce rates, so that by 2000, nearly half of all marriages ended in divorce in countries such as the United States, France, Germany, and Australia. See Goodin et al., *Discretionary Time*, 76; Bianchi et al., *Changing Rhythms of American Family Life*, 67.

19 Bianchi et al., *Changing Rhythms of American Family Life*; Lyn Craig, "Does Father Care Mean Fathers Share? A Comparison of How Mothers and Fathers in Intact Families Spend Time with Children," *Gender & Society* 20, no. 2 (2006): 259–81; Lyn Craig, *Contemporary Motherhood: The Impact of Children on Adult Time* (Aldershot, UK: Ashgate, 2007).

20 Bianchi et al., *Changing Rhythms of American Family Life*, 133–35.

21 According to the latest Pew report by Kim Parker and Wendy Wang, "Among those with children under the age of 18, 40% of working mothers and 34% of working fathers say they *always* feel rushed" (*Modern Parenthood*, 3; original emphasis).

22 Ibid. By comparison, lesbian families have a high level of shared housework and child care. See, for example, Timothy Biblarz and Evren Savci, "Lesbian, Gay, Bisexual, and Transgender Families," *Journal of Marriage and Family* 72, no. 3 (2010): 480–97.

23 Janet Gornick and Marcia Meyers, *Families that Work: Policies for Reconciling Parenthood and Employment* (New York: Russell Sage, 2003).

24 Bianchi et al., *Changing Rhythms of American Family Life*.

25 David Maume, "Reconsidering the Temporal Increase in Fathers' Time with Children," *Journal of Family and Economic Issues* 32, no. 3 (2011): 411–23. While fathers have nearly tripled their time with children since 1965, there is still a large gender gap in time spent with children: an average of seven hours per week for fathers, compared with fourteen hours per week for mothers. See Kim Parker and Wendy Wang *Modern Parenthood*, 3.

26 Bianchi et al., *Changing Rhythms of American Family Life*, 139.

27 Madeleine Bunting, *Willing Slaves: How the Overwork Culture is Ruling Our Lives* (London: Harper Collins, 2004).

28 Pierre Bourdieu, *Distinction: A Social Critique of the Judgement of Taste* (Cambridge, MA: Harvard University Press, 1984); Daniel Miller, *A Theory of Shopping* (Cambridge: Polity Press, 1998).

29 Manfred Garhammer, "Pace of Life and Enjoyment in Life," *Journal of Happiness Studies* 3, no. 3 (2002): 217–56.

30 Oriel Sullivan, "Busyness, Status Distinction and Consumption Strategies of the Income-Rich, Time-Poor," *Time & Society* 17, no. 1 (2008): 10.

31 Gershuny, "Busyness as the Badge of Honor"; Thorstein Veblen, *The Theory of the Leisure Class* (New York: Penguin Books, 1899). See also Luc Boltanski and Eve Chiapello, *The New Spirit of Capitalism* (London: Verso, 2007), who, from a very different perspective, argue that high prestige is placed on those who are constantly active. For an account that traces the culture of busyness to an even earlier period, see Benjamin Snyder, "From Vigilance to Busyness: A Neo-Weberian Approach to Clock Time," *Sociological Theory* 31, no. 3 (2013): 243–66.

32 Staffan Linder, *The Harried Leisure Class* (New York: Columbia University Press, 1970).

33 Gershuny, "Busyness as the Badge of Honor," 289.

34 Judy Wajcman, *Managing Like a Man: Women and Men in Corporate Management* (University Park: Penn State University Press, 1998); Judy Wajcman and Bill Martin, "Narratives of Identity in Modern Management: The Corrosion of Gender Difference," *Sociology* 36, no. 4 (2002): 985–1002.

35 Sullivan, "Busyness, Status Distinction and Consumption Strategies."

36 Ibid.; Tally Katz-Gerro and Oriel Sullivan, "Voracious Cultural Participation: Reinforcement of Gender and Social Status," *Time & Society* 19, no. 2 (2010): 193–219.

37 Dale Southerton and Mark Tomlinson, "'Pressed for Time'—The Differential Impacts of a 'Time Squeeze,'" *Sociological Review* 53, no. 2 (2005),: 232–33; Dale Southerton, "Analysing the Temporal Organization of Daily Life: Social Constraints, Practices and Their Allocation," *Sociology* 40, no. 3 (2006): 435–54.

38 See Bianchi et al., *Changing Rhythms of American Family Life*, appendix A, for a discussion of the reliability and validity of the time-diary approach.

39 Southerton, "Analysing the Temporal Organization of Daily Life."

40 Alan Warde, "Convenience Food: Space and Timing," *British Food Journal* 101, no. 7 (1999): 518–27.

41 Ibid., 518.

42 Laurent Lesnard, "Off-Scheduling within Dual-Earner Couples: An Unequal and Negative Externality for Family Time," *American Journal of Science* 114, no. 2 (2008): 466.

43 Southerton and Tomlinson, "Pressed for Time," 235.

44 Sullivan, "Time Waits for No (Wo)man"; Hochschild, *The Time Bind*.

45 Nancy Fraser, *Justice Interruptus: Critical Reflections on the "Postsocialist" Condition* (New York: Routledge, 1997), 26.

46 Michael Bittman and Judy Wajcman, "The Rush Hour: The Character of Leisure Time and Gender Equality," *Social Forces* 79, no. 1 (2000): 165–89.

47 Bianchi et al., *Changing Rhythms of American Family Life*, 98.

48 This research was conducted with Michael Bittman. See Bittman and Wajcman, "The Rush Hour."

49 Bianchi et al., *Changing Rhythms of American Family Life*, 85.

50 Replicating our study using US data, Mattingly and Bianchi, "Gender Differences in the Quantity and Quality of Free Time: The US Experience," did not find strong evidence that leisure time has become more fragmented for married mothers. However, they did find that pure free time (without housework, child care or grooming) fell from thirty-three to twenty-six hours for married mothers. Married fathers' leisure experience is more stable.

51 The results of the more detailed analysis of leisure show that the social cleavage between parents and nonparents is as important as gender differences, especially when children are young. Nevertheless, we show that

women have less adult leisure than comparable men, and that they are further disadvantaged by their disproportionate responsibility for the physical care of children (Bittman and Wajcman, "The Rush Hour," 185).

52 Michelle Budig and Nancy Folbre, "Activity, Proximity, or Responsibility? Measuring Parental Childcare Time," in Folbre and Bittman, *Family Time*, 63.

53 Suzzane Bianchi, "Maternal Employment and Time with Children: Dramatic Changes of Surprising Continuity?," *Demography* 37, no. 4 (2000): 401–14; Craig, *Contemporary Motherhood*.

54 Hartmut Rosa and William Scheuerman, eds., *High-Speed Society: Social Acceleration, Power, and Modernity* (University Park: Penn State University Press, 2009).

55 Zygmunt Bauman, *Liquid Life* (Cambridge: Polity Press, 2005), 84.

Chapter Four

1 Noelle Chesley, "Blurring Boundaries? Linking Technology Use, Spillover, Individual Distress, and Family Satisfaction," *Journal of Marriage and Family* 67, no. 5 (2005): 1237–48; Linda Duxbury and Rob Smart, "The 'Myth of Separate Worlds': An Exploration of How Mobile Technology has Redefined Work-Life Balance," *Creating Balance?*, ed. in Dans Kaiser et al. (Berlin: Springer-Verlag, 2011); Melissa Mazmanian, JoAnne Yates, and Wanda Orlikowski, "Ubiquitous Email: Individual Experiences and Organizational Consequences of BlackBerry Use" (paper presented at the 65th Annual Meeting of the Academy of Management, Honolulu, Hawaii, August 5–10, 2005).

2 Harry Braverman, *Labor and Monopoly Capital: The Degradation of Work in the Twentieth Century* (New York: Monthly Review Press, 1974).

3 Donald MacKenzie and Judy Wajcman,(eds., *The Social Shaping of Technology: How the Refrigerator Got Its Hum* (Milton Keynes, UK: Open University Press, 1985).

4 Vicki Smith, "Braverman's Legacy: The Labor Process Tradition at 20," *Work and Occupations* 21, no. 4 (1994): 431–21; Judy Wajcman, "New Connections: Social Studies of Science and Technology and Studies of Work," *Work, Employment & Society* 20, no. 4 (2006): 773–86.

5 David Noble, "Social Choice in Machine Design: The Case of Automatically Controlled Machine Tools," in *Case Studies in the Labor Process*, ed. Andrew Zimbalist (New York: Monthly Review Press, 1979), 19. His classic book is *Forces of Production: A Social History of Industrial Automation* (Oxford: Oxford University Press, 1984).

6 Nelly Oudshoorn and Trevor Pinch, eds., *How Users Matter: The Co-Construction of Users and Technology* (Cambridge, MA: MIT Press, 2005).

7 Figures taken from Lee Rainie and Barry Wellman, *Networked: The New Social Operating System* (Cambridge, MA: MIT Press, 2012), 174–75.

8 Robert Gordon, "Is US Economic Growth Over? Faltering Innovation Confronts the Six Headwinds," *CEPR Policy Insight* 63 (2012): 1–13.

9 Manuel Castells, *The Rise of the Network Society*, 2nd ed. (Malden, MA: Wiley-Blackwell, 2010), xxiv.

10 Ibid., chapter 4.

11 Arne L. Kalleberg, *Good Jobs, Bad Jobs: The Rise of Polarized and Precarious Employment Systems in the United States, 1970s–2000s* (New York: Russell Sage, 2011).

12 Mary Madden and Sydney Jones, "Networked Workers: Most Workers Use the Internet or Email at Their Jobs, but They Say These Technologies Are a Mixed Blessing for Them," *Pew Internet & American Life Project*, September 24, 2008. http://pewinternet.org/Reports/2008/Networked-workers.aspx.

13 Center for the Digital Future, *The Digital Future Project 2013—Surveying the Digital Future Year Eleven* (Los Angeles: University of Southern California, 2013). http://www.digitalcenter.org/wp-content/uploads/2013/06/2013-Report.pdf.

14 Pablo Boczkowski, *News at Work: Imitation in an Age of Information Abundance* (Chicago: Chicago University Press: 2010). On "cyberslacking," see Terrance Weatherbee, "Counterproductive Use of Technology at Work: Information & Communications Technologies and Cyberdeviancy, *Human Resource Management Review* 20 (2010): 35–44.

15 Madden and Jones, "Networked Workers."

16 Noelle Chesley, Andra Siibak, and Judy Wajcman, "Information and Communication Technology Use and Work-Life Integration," in *Handbook of Work-Life Integration of Professionals: Challenges and Opportunities*, ed. Debra Major and Ronald Burke (Cheltenham, UK: Elgar Publications, 2013).

17 An exception is David J. Maume and David A. Purcell, "The 'Over-Paced American: Recent Trends in the Intensification of Work," *Research in the Sociology of Work* 17 (2007): 252–83.

18 Francis Green, *Demanding Work. The Paradox of Job Quality in the Affluent Economy* (Princeton, NJ: Princeton University Press, 2006).

19 Alan Felstead, Duncan Gallie, Francis Green, and Hande Inanc, *Work Intensification in Britain: First Findings from the Skills and Employment Survey 2012* (London: Centre for Learning and Life Chances in Knowledge Economies and Society, Institute of Education, 2013).

20 Francis Green, "Why Has Work Effort Become More Intense?," *Industrial Relations* 43, no. 4 (2004): 709. In a similar vein, Tali Kristal argues that the decline of US workers' share of national income is not a direct result of computer-based technologies (as put forward by economists), but rather that computerization has reduced labor's share *indirectly* through its role in eroding the power of labor unions ("The Capitalist Machine: Computerization, Workers' Power, and the Decline in Labor's Share within U.S. Industries," *American Sociological Review* 78, no. 3 [2013]: 361–89).

21 Michael Bittman, Judith Brown, and Judy Wajcman, "The Mobile Phone, Perpetual Contact and Time Pressure," *Work, Employment and Society* 23,

no. 4 (2009): 673–91. See also Noelle Chesley, "Information and Communication Technology Use, Work Intensification, and Employee Strain and Distress," *Work, Employment & Society* (forthcoming).

22 Jane Wakefield, "Turn Off E-Mail and Do Some Work," *BBC News*, October 19, 2007. http://news.bbc.co.uk/go/pr/fr/-/1/hi/technology/7049275.stm.

23 Stephen Barley, Debra Meyerson, and Stine Grodal, "Email as a Source and Symbol of Stress," *Organization Science* 22, no. 4 (2011): 887.

24 Ibid., See also Paul Leonardi and Stephen Barley, "What's Under Construction Here? Social Action, Materiality, and Power in Constructivist Studies of Technology and Organizing," *Academy of Management Annals* 4, no. 1 (2010): 1–52, for an overview of the increasing influence of STS on organization studies.

25 Barley et al., "Email as a Source and Symbol of Stress," 903.

26 For example, Sylvia Ann Hewlett, "Is Your Extreme Job Killing You?," *Harvard Business Review*, August 22, 2007.

27 Brid O'Conaill and David Frohlich, "Timespace in the Workplace: Dealing with Interruptions," *Companion Proceedings of ACM CHI 1995* (New York: ACM CHI Companion, 1995), 262. See also Quintus Jett and Jennifer George, "Work Interrupted: A Closer Look at the Role of Interruptions in Organizational Life," *Academy of Management Review* 28, no. 3 (2003): 498–507; Mary Czerwinski, Eric Horvitz, and Susan Wilhite, "A Diary Study of Task Switching and Interruptions," *Companion Proceedings of ACM CHI 2004* (New York: ACM CHI Companion, 2004), 175–82; Mark Ellwood, "Time Priorities for Top Managers" (paper presented at the International Association of Time Use Researchers [IATUR] Conference, Halifax, Nova Scotia, November 2–4, 2005); Victor Gonzalez and Gloria Mark, "'Constant, Constant, Multi-Tasking Craziness': Managing Multiple Working Spheres," *Companion Proceedings of ACM CHI 2004* (New York: ACM CHI Companion, 2004): 113–20.

28 My own approach has much in common with Wanda Orlikowski's practice-based approach: "*every* organizational practice is *always* bound with materiality" ("Sociomaterial Practices: Exploring Technology at Work," *Organization Studies* 28, no. 9 [2007]: 1436; original emphasis). See also Lucy Suchman's *Human-Machine Reconfigurations: Plans and Situated Actions, Second Edition* (New York: Cambridge University Press, 2007), 286: "The point in the end is not to assign agency either to persons or to things but to identify the materialization of subjects, objects, and the relations between them as an effect . . . of ongoing sociomaterial practices."

29 The participants in our sample, who attend to general managerial functions, finance and accounting, organizational development and strategic affairs, and human resources, are typical of those who focus on information, and, as such, I broadly refer to them as "knowledge workers." Details of the study can be found in Judy Wajcman and Emily Rose, "Constant Connectivity: Rethinking Interruptions at Work," *Organization Studies* 32, no. 7 (2011): 941–62.

30 If the telephone rings and an employee responded to this, we recorded this as the telephone being the reason for the change in work activity. Alternatively, if someone was alerted to a message on his or her telephone and did not respond immediately, instead checking the subsequent voice message five minutes later, this change in work activity is categorized under "self-initiated," as the worker decided when to change his or her work activity to attend to the voice message.

31 Henry Mintzberg, *The Nature of Managerial Work* (New York: Harper & Row, 1973).

32 Melissa Mazmanian, Wanda Orlikowski, and JoAnne Yates, "The Autonomy Paradox: The Implications of Mobile Email Devices for Knowledge Professionals," *Organization Science* 24, no. 5 (2013): 1337–57.

33 How much this depends on the context and type of occupation and industry involved is highlighted by one of the authors in another study that shows clearly how, even within one organization, people enact technologies differently. See Melissa Mazmanian, "Avoiding the Trap of Constant Connectivity: When Congruent Frames Allow for Heterogeneous Practices," *Academy of Management Journal* 56, no. 5 (2013): 1225–50.

34 Mazmanian, Orlikowski, and Yates, "The Autonomy Paradox," 1353.

35 Steven Appelbaum, Adam Marchionni, and Arturo Fernandez, "The Multi-Tasking Paradox: Perceptions, Problems and Strategies," *Management Decision* 46, no. 9 (2008): 1313–25; Noelle Chesley, "Stay-at-Home Fathers and Breadwinning Mothers: Gender, Couple Dynamics, and Social Change," *Gender & Society* 25, no. 5 (2011): 642–64; Larry Rosen, *iDisorder: Understanding our Obsession with Technology and Overcoming its Hold on Us* (New York: Palgrave, 2012); Victor Gonzalez and Gloria Mark, "Constant, Constant, Multi-Tasking Craziness"; Norman Su and Gloria Mark, "Communication Chains and Multitasking," *Companion Proceedings of ACM CHI 2008* (New York: ACM CHI Companion, 2008), 262–63.

36 Eyal Ophir, Clifford Nass, and Anthony Wagner, "Cognitive Control in Media Multitaskers," *Proceedings of the National Academy of Sciences USA* 106, no. 37 (2009): 15583–87.

37 Rosen, *iDisorder*.

38 Rainie and Wellman, *Networked*, 273 (original emphasis).

39 Edward T. Hall, *The Dance of Life: The Other Dimension of Time* (New York: Anchor Books, 1989), 52. See Valerie Bryson, *Gender and the Politics of Time: Feminist Theory and Contemporary Debates* (Bristol, UK: Policy Press, 2007), 130, for an excellent feminist discussion of these temporalities.

40 Allen Bluedorn, *The Human Organization of Time* (Stanford, CA: Stanford University Press, 2002), 62.

41 Ibid.

42 Noelle Chesley, "Technology Use and Employee Assessments of Work Effectiveness, Workload, and Pace of Life," *Information, Communication & Society* 13, no. 4 (2010): 485–514.

43 Shira Offer and Barbara Schneider, "Revisiting the Gender Gap in Time-

Use Patterns: Multitasking and Well-Being among Mothers and Fathers in Dual-Earner Families," *American Sociological Review* 76, no. 6 (2011): 828.

44 Claudio Ciborra and Ole Hanseth, "From tool to Gestell: Agendas for Managing the Information Infrastructure," *Information Technology & People* 11, no. 4 (1998): 321–22. See also the Infrastructure Series, published by MIT Press and edited by Geoffrey Bowker and Paul Edwards.

45 Richard Sennett, *The Corrosion of Character: The Personal Consequences of Work in the New Capitalism* (New York: W. W. Norton & Company, 1998).

46 For example, Sennett's gender blindness. See Judy Wajcman and Bill Martin, "Narratives of Identity in Modern Management: The Corrosion of Gender Difference?," *Sociology* 36, no. 4 (2002): 985–1002.

47 Mary Madden and Sydney Jones, "Networked Workers."

Chapter Five

1 Ruth Schwartz Cowan, "The 'Industrial Revolution' in the Home: Household Technology and Social Change in the 20th Century," *Technology and Culture* 17, no. 1 (1976): 8–9.

2 Ha Joon Chang, *23 Things They Don't Tell You about Capitalism* (London: Allen Lane, 2010).

3 Jonathan Gershuny, *Changing Times: Work and Leisure in Postindustrial Society* (Oxford: Oxford University Press, 2003), chapter 7.

4 Talcott Parsons and Robert Freed Bales, *Family, Socialization and Interaction Process* (London: Routledge, 1956).

5 Ruth Schwartz Cowan, *More Work for Mother: The Ironies of Household Technology from the Open Hearth to the Microwave* (New York: Basic Books, 1983).

6 Susan Strasser, *Never Done: A History of American Housework* (New York: Pantheon Books, 1982).

7 According to the BaseKit UK survey, in 2012, women spent on average 3.3 hours per week shopping (2.3 hours online and 1 hour in store), compared to 2.9 hours per week for men (2.1 hours online and 0.8 hours in store).

8 Cowan, *More Work for Mother*, 100.

9 Ibid., 99.

10 Barbara Ehrenreich and Deirdre English, *For Her Own Good: 150 Years of Experts Advice to Women* (London: Pluto Press, 1979).

11 Ruth Schwartz Cowan, "From Virginia Dare to Virginia Slims: Women and Invention in America," *Technology and Culture* 20, no. 1 (1979): 59.

12 See Judy Wajcman, *Feminism Confronts Technology* (Cambridge: Polity Press, 1991), chapter 4.

13 In the UK and the United States, for example, men's overall contribution to domestic work increased from 90 and 105 minutes per day, respectively, in the 1960s to 148 and 173 minutes per day, respectively, in the early 2000s. Women's domestic work has gradually declined over this period but still amounts to 280 minutes per day in the UK and 272 minutes per day in the

United States. See Man Yee Kan, Oriel Sullivan, and Jonathan Gershuny, "Gender Convergence in Domestic Work: Discerning the Effects of Interactional and Institutional Barriers from Large-Scale Data," *Sociology* 45, 2 (2011): 237.

14 Ibid.

15 Ibid., 240.

16 Michael Bittman, James Rice, and Judy Wajcman, "Appliances and Their Impact: The Ownership of Domestic Technology and Time Spent on Housework," *British Journal of Sociology* 55, no. 2 (2004): 400–23.

17 The figures in this paragraph are all sourced from the Department of Energy and Climate Change, "Statistics at the DECC," available on the Gov. UK website: https://www.gov.uk/government/organisations/department-of-energy-climate-change/about/statistics.

18 Elizabeth Shove, *Comfort, Cleanliness and Convenience: The Social Organization of Normality* (London: Berg, 2003), 90.

19 For example, a survey by Unilever found that the average length of showers in the UK is eight minutes (Mark Kinver, "People's Showering Habits Revealed in Survey," BBC News, November 22, 2011. http://www.bbc.co.uk/news/science-environment-15836433).

20 Alan Warde, "Convenience Food: Space and Timing," *British Food Journal* 101, no. 7 (1999): 518–27.

21 Dale Southerton, "Ordinary and Distinctive Kitchens; or a Kitchen Is a Kitchen Is a Kitchen," in *Ordinary Consumption*, ed. Jukka Groncow and Alan Warde (London: Routledge, 2001).

22 Shove, *Comfort, Cleanliness and Convenience*, 182.

23 Current statistics on food expenditure at home and away from home in the United States are available at the USDA Economic Research service website: http://www.ers.usda.gov/data-products/food-expenditures.aspx#.UanGvkBqkp8.

24 Sanjiv Gupta, "Autonomy, Dependence, or Display? The Relationship between Married Women's Earnings and Housework," *Journal of Marriage and Family* 69, no. 2 (2007): 413.

25 Bridget Anderson, *Doing the Dirty Work? The Global Politics of Domestic Labour* (London: Zed Books, 2000); Barbara Ehrenreich and Arlie Hochschild, eds., *Global Woman: Nannies, Maids and Sex Workers in the New Economy* (New York: Metropolitan Books, 2003), 1–14.

26 Nancy Folbre, *Valuing Children: Rethinking the Economics of the Family* (Cambridge, MA: Harvard University Press, 2007).

27 Annette Lareau, *Unequal Childhoods: Class, Race, and Family Life* (Berkeley: University of California Press, 2003).

28 Lyn Craig and Killian Mullan, "How Mothers and Fathers Share Childcare: A Cross-National Time-Use Comparison," *American Sociological Review* 76, no. 6 (2011): 834–61.

29 Ibid., 847.

30 Laurent Lesnard, "Off-Scheduling within Dual-Earner Couples: An Un-

equal and Negative Externality for Family Time," *American Journal of Science* 114, no. 2 (2008): 474.

31 Miriam Glucksmann, *Cottons and Casuals: The Gendered Organisation of Labour in Time and Space* (Durham, UK: Sociology Press, 2000), 114.

32 Karen Davies, "The Tensions between Process Time and Clock Time in Care-Work: The Example of Day Nurseries," *Time & Society* 3, no. 3 (1994): 277–303; see also Barbara Adam, *Timewatch: The Social Analysis of Time* (Cambridge: Polity Press, 1995).

33 Julia Kristeva, "Women's Time," *Signs* 7, no. 1 (1981): 13–35; Valerie Bryson, *Gender and the Politics of Time: Feminist Theory and Contemporary Debates* (Bristol, UK: Policy Press, 2007), 142.

34 Paul Dourish and Genevieve Bell, *Divining a Digital Future: Mess and Mythology in Ubiquitous Computing* (Cambridge, MA: MIT Press, 2011), 165.

35 Sherry Turkle, *Life on the Screen: Identity in the Age of the Internet* (New York: Touchstone, 1997); Turkle, *The Second Self: Computers and the Human Spirit* (Cambridge, MA: MIT Press, 2005), Turkle, *Alone Together: Why We Expect More from Technology and Less from Each Other* (New York: Basic Books, 2011).

36 Turkle, *Alone Together*, 139.

37 Avner Offer, *The Challenge of Affluence: Self-Control and Well-Being in the United States and Britain since 1950* (Oxford: Oxford University Press, 2006), 187.

Chapter Six

1 Tony Paterson, "Out of the Office and Not Taking Emails: Victory for VW Workers," *Independent*, December 24, 2011.

2 Anthony Giddens, *Modernity and Self-Identity* (Cambridge: Polity Press, 1991); Christopher Lasch, *Haven in a Heartless World* (New York: Basic Books, 1977); Eli Zaretsky, *Capitalism, the Family and Personal Life* (New York: Harper & Row, 1976).

3 Mimi Sheller and John Urry, "Mobile Transformations of 'Public' and 'Private' Life," *Theory, Culture & Society* 20, no. 3 (2003): 122.

4 Jon Agar, *Constant Touch: A Global History of the Mobile Phone* (Cambridge: Icon Books, 2003); James Katz and Mark Aakhus, eds., *Perpetual Contact: Mobile Communication, Private Talk, Public Performance* (Cambridge: Cambridge University Press, 2002); Christian Licoppe, "'Connected Presence': The Emergence of a New Repertoire for Managing Social Relationships in a Changing Communication Technoscape," *Environment and Planning D: Society and Space* 22, no. 1 (2004): 135–56.

5 See, for example, Gerard Goggin, *New Technologies and the Media* (New York: Palgrave Macmillan, 2012); Nicola Green and Leslie Haddon, *Mobile Communications: An Introduction to New Media* (Oxford: Berg, 2009); Nick Couldry, *Media, Society, World: Social Theory and Digital Media Practice* (Cambridge: Polity Press, 2012).

6 Mary Madden and Sydney Jones, "Networked Workers," *Pew Internet & American Life Project*, September 24, 2008. http://pewinternet.org/Reports/2008/Networked-workers.aspx.

7 Linda Duxbury, Ian Towers, Christopher Higgins, and John Thomas, "From 9 to 5 to 24 and 7: How Technology Redefined the Work Day," in *Information Resources Management: Global Challenges*, ed. Wai Law (Hershey, PA: Idea Group Publishing, 2006).

8 Noelle Chesley, "Blurring Boundaries? Linking Technology Use, Spillover, Individual Distress, and Family Satisfaction," *Journal of Marriage and Family* 67, no. 5 (2005): 1237–48.

9 Morgan, David, "Risk and Family Practices: Accounting for Change and Fluidity in Family Life," in *The New Family?*, ed. Elizabeth Silva and Carol Smart (London: Sage, 1999), 20. See also Lynn Jamieson, *Intimacy: Personal Relationships in Modern Societies* (Cambridge: Polity Press, 1998).

10 I have discussed elsewhere how Morgan's conceptual ploy of treating the family as a verb has striking parallels with actor-network theory, in which nonhuman objects are conceived of as actants and society is viewed as a *doing* rather than a *being*. See Judy Wajcman, Michael Bittman, and Jude Brown, "Intimate Connections: The Impact of the Mobile Phone on Work/Life Boundaries," in *Mobile Technologies: From Telecommunications to Media, ed.* Gerard Goggin and Larissa Hjorth (London: Routledge, 2008).

11 Peter Glotz, Stefan Bertschi, and Chris Locke, eds., *Thumb Culture: The Meaning of Mobile Phones for Society* (New Brunswick, NJ: Transaction Publishers, 2005); James Katz, ed., *Machines that Become Us* (New Brunswick, NJ: Transaction Publishers, 2003); Katz and Aakhus, *Perpetual Contact*.

12 See Judy Wajcman, Michael Bittman, and Jude Brown, "Families without Borders: Mobile Phones, Connectedness and Work-Home Divisions," *Sociology* 42, no. 4 (2008): 635–52; and Michael Bittman, Jude Brown, and Judy Wajcman, "The Mobile Phone, Perpetual Contact and Time Pressure," *Work, Employment and Society* 23, no. 4 (2009): 673–79, for full details of this study. Although the research predates the heavy use of smartphones, it is still worth reporting as it is still a relatively rare, nationally representative survey combined with both interview and diary data.

13 This was confirmed by our analysis of the phone log data using the standard "family strains and gains scale." Contrary to expectations, the number of calls made and received on a mobile phone is *not* significantly associated with increased work to family spillover (or work-family strain). It seems that job characteristics have a far greater influence on work-family spillover than mobile communications—especially work stress and if employees are working longer than their preferred working hours.

14 Rich Ling, *The Mobile Connection* (San Francisco: Morgan Kaufmann, 2004); Ling, *New Tech, New Ties: How Mobile Communication is Reshaping Social Cohesion* (Cambridge, MA: MIT Press, 2008).

15 Emily Rose, "Access Denied: Employee Control of Personal Communications at Work," *Work, Employment & Society* 27, no. 4 (2013): 694–710. In

another study, Noelle Chesley also found that connecting with family and friends while at work can reduce the likelihood that an employee reports an increase in work stress. See "Information and Communication Technology Use, Work intensification, and Employee Strain and Distress," *Work, Employment and Society* (forthcoming).

16 Ling, *The Mobile Connection*, 69.

17 Lewis Mumford, *Technics and Civilization* (New York: Harcourt, Brace, 1934), 14.

18 This point is well made by Green and Haddon, *Mobile Communications*, 83.

19 Licoppe, "Connected Presence."

20 For a detailed account of how much time is spent throughout the day on media activity, see Ofcom, "Communications Market Report," August 19, 2010, 31 (fig. 1.13). http://stakeholders.ofcom.org.uk/market-data-research/market-data/communications-market-reports/cmr10/uk/.

21 Sherry Turkle, *Alone Together: Why We Expect More From Technology And Less From Each Other* (New York: Basic Books, 2011).

22 See, for example, Barry Wellman, Anabel Quan-Hasse, James Witte, and Keith Hampton, "Does the Internet Increase, Decrease, or Supplement Social Capital? Social Network, Participation and Community Commitment," *American Behavioral Scientist* 45, no. 4 (2001): 437–56.

23 William Dutton and Grant Blank, "Next Generation Users: The Internet in Britain," *Oxford Internet Survey* (Oxford: Oxford Internet Institute, 2011). http://oxis.oii.ox.ac.uk/.

24 Ofcom, "The Communications Market Report: United Kingdom: The Reinvention of the 1950s Living Room," August 2013, 69. http://stakeholders.ofcom.org.uk/market-data-research/market-data/communications-market-reports/cmr13/uk/.

25 Barry Wellman, Aaron Smith, Amy Wells, and Tracy Kennedy, "Networked Families," *Pew Internet & American Life Project*, October 19, 2008. http://www.pewinternet.org/2008/10/19/networked-families/.

26 Lana Rakow, *Gender on the Line: Women, the Telephone, and Community Life* (Urbana: University of Illinois Press, 1992).

27 Giddens, *Modernity and Self-Identity*; Giddens, *The Transformation of Intimacy: Sexuality, Love and Eroticism in Modern Societies* (Cambridge: Polity Press, 1992). See also Lynn Jamieson, "Intimacy Transformed? A Critical Look at the 'Pure Relationship,'" *Sociology* 33, no. 3 (1999): 477–94.

28 For an excellent overview of these issues, see Lynn Jamieson, "Personal Relationships, Intimacy and the Self in a Mediated and Global Digital Age" in *Digital Sociology: Critical Perspectives*, ed. Kate Orton-Johnson and Nick Prior (London: Palgrave Macmillan, 2013).

29 Sonia Livingstone, *Children and the Internet* (Cambridge: Polity Press, 2009); Gill Valentine, "Globalizing Intimacy: The Role of Information and Communication Technologies in Maintaining and Creating Relationships," *Women's Studies Quarterly* 34, no. 1/2 (2006): 367.

30 Kennedy et al., "Networked Families."

31 Figures taken from Mizuko Ito et al., *Hanging Out, Messing Around, and Geeking Out: Kids Living and Learning with New Media* (Cambridge, MA: MIT Press, 2010); and Lee Rainie and Barry Wellman, *Networked: The New Social Operating System* (Cambridge, MA: MIT Press, 2012).

32 Rich Ling and Brigitte Ytrri, "Control, Emancipation, and Status: The Mobile Telephone in Teens' Parental and Peer Relationships," in *Computers, Phones and the Internet: Domesticating Information Technology*, ed. Robert Kraut, Malcolm Brynin and Sara Kiesler (Oxford: Oxford University Press, 2006).

33 Ibid., 228.

34 Ito et al., *Hanging Out, Messing Around, and Geeking Out.*

35 See Green and Haddon, *Mobile Communications*, 97, and Nancy Baym, *Personal Connections in the Digital Age* (Cambridge: Polity Press, 2010), for an overview of studies on mobile phone and new media usage.

36 Richard Harper, *Texture: Human Expression in the Age of Communications Overload* (Cambridge, MA: MIT Press, 2010), 41.

37 Leslie Haddon and Jane Vincent, "Children's Broadening Use of Mobile Phones," in *Mobile Technologies: From Telecommunications to Media*, edited by Gerard Goggin and Larissa Hjorth, 37–49 (Abingdon, UK: Routledge, 2009).

38 See Nielsen, "Television Audience Report 2010–2011." http://www.nielsen.com/us/en/reports/2011/television-audience-report-2010-2011.html.

39 Olle Findahl, "How to Measure the Use of the Internet? A Comparison between Questionnaires, Dairies and Traffic Measurements" (paper presented at the World Internet Institute Conference, Lisbon, Portugal, July 7, 2010).

40 John Robinson, "IT Use and Leisure Time Displacement: Convergent Evidence over the Last 15 Years," *Information, Communication & Society* 14, no. 4 (2011): 507.

41 *The New Multi-Screen World Study*, Google, August 2012. http://www.thinkwithgoogle.com/research-studies/the-new-multi-screen-world-study.html. According to Ofcom, *Communications Market Report*, just over half of all UK adults are regular media multitaskers.

42 "Web Saves Retailers from a Wet Christmas," *Financial Times*, January 8, 2013.

43 Tiziana Terranova, "Free Labor: Producing Culture for the Digital Economy, *Social Text* 18, no. 2 (2000): 33. See also Trebor Scholz, ed., *Digital Labor: The Internet as Playground and Factory* (New York: Routledge, 2013).

44 The sociological literature on consumer culture is now vast. See, for example, Dale Southerton, ed., *Encyclopedia of Consumer Culture* (Thousand Oaks, CA: Sage, 2011).

45 Jonathan Zittrain, *The Future of the Internet: And How to Stop It* (New Haven, CT: Yale University Press, 2008), 70.

46 Zittrain's real target is what he calls sterile, tethered systems like the

iPhone, whose functionality is locked by Apple, compared with generative PCs. While persuasive at the time, subsequent developments, such as the explosion of creativity around writing software applications, or apps, for the iPhone and iPad, tell a rather different story. Apple's decision to open its devices turned out to be a brilliant commercial move, especially as they retain 30 percent of each app sale. Nonetheless, Zittrain's point about the significance of generativity still holds.

47 "As Boom Lures App Creators, Tough Part Is Making a Living," *New York Times*, November 17, 2012.

48 Yochai Benkler, *The Wealth of Networks: How Social Production Transforms Markets and Freedom* (New Haven, CT: Yale University Press, 2006), 275.

49 See, for example, Manuel Castells, *The Internet Galaxy* (New York: Oxford University Press, 2001), 374.

Chapter Seven

1 Robert Skidelsky and Edward Skidelsky, *How Much Is Enough? The Love of Money, and the Case for the Good Life* (London: Allen Lane, 2012), 16.

2 Ben Agger, *Speeding Up Fast Capitalism* (Boulder CO: Paradigm, 2004), 157.

3. More can be read about this on the Qualified Self movement website, at http://quantifiedself.com/.

4 Skidelsky and Skidelsky, *How Much Is Enough?*, 25.

5 Juliet Schor, *Plenitude: The New Economics of True Wealth* (New York: Penguin, 2010), 103. See also books by Benjamin Kline Hunnicutt, most recently, *Free Time: The Forgotten American Dream* (Philadelphia, PA: Temple University Press, 2013).

6 They also note that working hours vary significantly between countries, with the United States and the UK being outliers with regard to long hours and income inequality, compared to, say, Sweden and the Netherlands.

7 See, for example, the writings of André Gorz.

8 Bertrand Russell, *In Praise of Idleness* (London: Allen & Unwin, 1935); Paul Lafargue, *The Right to be Lazy and Other Studies* (Chicago: Charles H. Kerr and Company, 1883).

9 Kathi Weeks, *The Problem with Work: Feminism, Marxism, Antiwork Politics and Postwork Imaginaries* (Durham, NC: Duke University Press, 2011), 172.

10 Jeremy Rifkin, *Time Wars: The Primary Conflict in Human History* (New York: Henry Holt and Company, 1987), 197.

11 Jonathan Gershuny, *After Industrial Society? The Emerging Self-service Economy* (London: Macmillan, 1978).

12 Hilary Stout, "For Shoppers, Next Level of Instant Gratification," *New York Times* October 8, 2013. Intriguingly, the man who is bringing us instantaneity has donated $42 million to the construction of the Clock of the Long Now, which is supposed to tick for ten thousand years (see Brad Stone, *The Everything Store: Jeff Bezos and the Age of Amazon* [London: Bantam, 2013]).

13 Ben Agger, "iTime: Labor and Life in a Smartphone Era," *Time & Society* 20, no. 1 (2011): 124.

14 John Tomlinson, *The Culture of Speed: The Coming of Immediacy* (London: Sage, 2007), 139.

15 "The Mind Business," *Financial Times*, August 24, 2012.

16 Slow Science website, http://slow-science.org/.

17 Wendy Parkins and Geoffrey Craig, *Slow Living* (Oxford: Berg, 2006), 42 (original emphasis). See also Helga Nowotny, *Time: The Modern and Postmodern Experience* (Cambridge: Polity Press, 2005).

18 Parkins and Craig, *Slow Living*, 53. My account of the slow food movement, an extensive organization of some eighty thousand members across a hundred countries, relies on this excellent book. See also Carl Honore, *In Praise of Slow: How a Worldwide Movement is Challenging the Cult of Speed* (London: Orion Books, 2005).

19 See Parkins and Craig, *Slow Living*. See also John Urry's elaboration of a slow-moving and ponderous "glacial time" in chapter 6 of *Sociology Beyond Societies* (London: Routledge, 2000).

20 Parkins and Craig, *Slow Living*, 140.

21 Susan Douglas, *Listening In: Radio and the American Imagination* (New York: Times Books, 1999), 21.

22 Miles Johnson, "Advance of the Robots," *Financial Times*, April 5, 2013.

23 Eric Schmidt and Jared Cohen, *The New Digital Age: Reshaping the Future of People, Nations and Business* (New York: Knopf, 2013).

24 A study by Roja Bandari, Sitaram Asur, and Bernardo Huberman found that technology content could predict to an accuracy of 84 percent the spread of news items on the web ("The Pulse of News in Social Media: Forecasting Popularity," paper presented at the sixth international Association of the Advancement of Artificial Intelligence Conference on Weblogs and Social Media, 2012).

25 See Evgeny Morozov, *To Save Everything, Click Here: Technology, Solutionism and the Urge to Fix Problems That Don't Exist* (New York: Allen Lane, 2013) for a critique of what he terms "technological solutionism."

26 Andrew Feenberg provides a good introduction to this literature, for example, *Transforming Technology: A Critical Theory Revisited* (Oxford: Oxford University Press, 2002).

27 Nik Brown and Mike Michael, "A Sociology of Expectations: Retrospecting Prospects and Prospecting Retrospects," *Technology Analysis and Strategic Management* 15, no. 1 (2003): 3–18. See also Paul Dourish and Genevieve Bell, *Divining A Digital Future: Mess and Mythology in Ubiquitous Computing* (Cambridge, MA: MIT Press, 2011) on the "proximate future."

28 Jennifer Karns Alexander, *The Mantra of Efficiency: From Waterwheel to Social Control* (Baltimore, MD: John Hopkins University Press, 2008), xi–xii. There is an extensive STS literature on how scientific standards function in the service of efficiency, erasing differences and precluding alternative

modes of operating in the world. See, for example, the writings of Susan Leigh Star, Geoff Bowker, Steve Epstein, and Stefan Timmermans.

29 This concept of efficiency certainly does not take into account the human, material, and environmental costs of ever more powerful technology. While consumers think of mobile phones as timesaving devices, for the Chinese workers who produce them and the Congolese slaves who extract the minerals required, they represent something entirely different.

30 Graeme Kirkpatrick, *Technology and Social Power* (Basingstoke, UK: Palgrave Macmillan, 2008), 142. See also Lucas Introna and Helen Nissenbaum, "Shaping the Web; Why the Politics of Search Engines Matters," *Information Society* 16, no. 3 (2000): 169–85.

31 Tarleton Gillespie, "The Relevance of Algorithms," in *Media Technologies: Essays on Communication, Materiality, and Society*, ed. Tarleton Gillespie, Pablo Boczkowski and Kirsten Foot (Cambridge, MA: MIT Press, 2014); Jaron Lanier, *Who Owns the Future?* (New York: Simon & Schuster, 2013).

32 Andrew Barry, *Political Machines: Governing a Technological Society* (London: Athlone Press, 2001), 212.

33 Jonathan Sterne, "Out with the Trash: On the Future of New Media," in *Residual Media*, ed. Charles Acland (Minneapolis: University of Minnesota Press, 2007), 28. By 2017, the annual volume of end-of-life TVs, mobile phones, computers, monitors, e-toys, and other electronic products is expected to be 65.4 million tons. Much toxic e-waste is dumped in the third world (see United Nations "Solving the E-Waste Problem [StEP] Initiative," 2013). http://isp.unu.edu/research/step/.

34 Andrew Barry, *Political Machines*, 211. See also Sheila Jasanoff, *Designs on Nature: Science and Democracy in Europe and the United States* (Princeton, NJ: Princeton University Press, 2005).

35 Judy Wajcman, *TechnoFeminism* (Cambridge: Polity Press, 2004), 111. See also various works on the masculinity of engineering by Wendy Faulkner, for example, "'Nuts and Bolts and People': Gender-Troubled Engineering Identities," *Social Studies of Science* 37, no. 3 (2007): 331–56.

36 Interestingly, an "engineering mindset" has recently been invoked in a sociological analysis about the preponderance of engineers among Islamic radicals. See Diego Gambetta and Steffen Hertog, "Why Are There So Many Engineers among Islamic Radicals?," *European Journal of Sociology* 50, no. 2 (2009): 201–30.

37 Morozov, *To Save Everything, Click Here*, 171.

38 See "Special Report on Cars," *Economist*, April 20, 2013. See also Kingsley Dennis and John Urry, *After the Car* (Cambridge: Polity Press, 2009).

39 See chapter 4 of Judy Wajcman, *Feminism Confronts Technology* (Cambridge: Polity Press, 1991), for my discussion of alternatives to individualized housework.

40 See Lucy Suchman's groundbreaking book *Human-Machine Reconfigurations: Plans and Situated Actions*, 2nd ed. (New York: Cambridge University

Press, 2007), 276; her literature overview, "Feminist STS and the Sciences of the Artificial," in *The Handbook of Science and Technology Studies,* 3rd ed., ed. Edward Hackett et al. (Cambridge, MA: MIT Press, 2008); and "Subject Objects," *Feminist Theory* 12, no. 2 (2011): 119–45. On technoscience as a "material-semiotic practice," see Donna Haraway, *Simians, Cyborgs, and Women: The Reinvention of Nature* (London: Routledge, 1990).

41 Clive Cookson, "Inside the MIT Media Lab," *Financial Times,* May 4, 2013. MIT is the interdisciplinary research and design center at the Massachusetts Institute of Technology. See Anne Balsamo, *Designing Culture: The Technological Imagination at Work* (Durham, NC: Duke University Press, 2011), for a unique vision of how we can gender the technological imagination.

42 Janet Roitman, *Anti-Crisis* (Durham, NC: Duke University Press, 2013).

Index